Understanding the Spirit World & Beyond

Understanding the Spirit World & Beyond

Fulfilling Mankind's Primary Purpose on Earth

Richard Gene, Ph.D.

Alive Book Publishing

Understanding the Spirit World and Beyond
Copyright © 2020 by Richard Gene, Ph.D.

All rights reserved.
No part of this book may be reproduced or transmitted in any form or by any means without written permission from the publisher and author.

Additional copies may be ordered from the publisher for educational, business, promotional or premium use. For information, contact ALIVE Book Publishing at: alivebookpublishing.com, or call (925) 837-7303.

Book Design by Alex Johnson

ISBN 13
978-1-63132-096-5

Library of Congress Control Number: 2020911963

Library of Congress Cataloging-in-Publication Data
is available upon request.

First Edition

Published in the United States of America by
ALIVE Book Publishing and ALIVE Publishing Group,
imprints of Advanced Publishing LLC
3200 A Danville Blvd., Suite 204, Alamo, California 94507
alivebookpublishing.com

PRINTED IN THE UNITED STATES OF AMERICA

10 9 8 7 6 5 4 3 2 1

Table of Contents

Preface..17

New Discoveries & Developments Made with the
Spiritual Model Presented in Reference 1.............................21

Special Acknowledgement and Dedication..........................27

Acknowledgements...29

Part One

The Spirit World, the Spiritual Model, and Mankind's Primary Purpose on Earth

Chapter One: Background Information and Examples
of New Discoveries & Developments......................................33

1.1.	Introduction..33
1.2.	The Source that Gives Life to Living Things...............36
1.3.	The Earliest Form of Recognizable Life on Earth Has to Be the Simplest Possible Living Thing..........................40
1.4.	Our Universe Is Huge and Is Filled with All Sizes of Galaxies, Stars, Planets, and Black Holes for a Reason..............40
1.5.	Functional MRI Studies of Brain Activity Patterns Support the Notion that the Brain Communicates with the Spirit..43
1.6.	The Way Things Work in the Spirit World Is Very Positive for All Religions and All Perceptions of God..................44

1.7. Brief Statement Regarding How to Fulfill Mankind's Primary Purpose for Being Here on Earth.................47
1.8. How Far Along Is Mankind in Fulfilling its Primary Purpose...50

Chapter Two: Our Spiritual Advancements Need to Match Our Technical Advancements........................53

2.1. It Is Time We Improve Mankind's Overall Behavior.............53
2.2. A New and More Effective Way to Pursue Spiritual Advancements...58
2.3. Interactions and Interplays Between the Spiritual Parts and the Technical Parts of Things.................60
2.4. Differences Between the Part of Life in the Spirit World and the Part of Life in Our Physical World..........62
2.5. A Civilization at Risk...68
2.6. A Democratic Governing System Is Fragile and Needs Complete Support...79

Chapter Three: Mankind's Current State of Human Nature...........83

3.1. Mankind's Human Nature Is Changeable........................83
    1. Creatures on the Galapagos Islands
    2. Domesticating animals
    3. Domesticated creatures released into the wild
3.2. A Leader's Poor Behavior Could Be Sustained by Our Current State of Human Nature.................................87
    1. They do not have anyone above them to prevent them from behaving poorly, especially if they are dictators
    2. The attitude and values of the population keep changing from one generation to the next
3.3. Long-Term Long-Range Thinking vs. Short-Term Short-Range Thinking...90
    1. Our current state of human nature would have us think short-term short-range much more often than long-term long-range

    2. Long-term long-range actions and short-term short-range actions could both be valid and yet be opposing
    3. A real-life example

3.4. Steering the Economy Is Like Steering a Huge Ocean Liner at Sea 92
3.5. Take a Partnership Attitude and not an Opponent Attitude 93
3.6. Our Actions Tend to Be Locked in by Our Institutions and They Do Not Match the Need of the Spirit World 95
3.7. The Urge to Fight or Flight 96
3.8. Some People in Government Behave Like an Adult Gang 100
3.9. Some People Need a Strong Leader Simply for the Sake of Having a Strong Leader 102
3.10. A Leader Needs to Understand the Nation's Population 104
3.11. Interactions with Intelligent Beings from Outer Space 105
3.12. The Possible Formation of a Military Space Force 107

Chapter Four: The Spirit World and the Spiritual Model Presented in Reference 1 109

4.1. The Spiritual Model Presented in Reference 1 109
    1. The indisputable initial concept
    2. Knowledge is generated by experiences
    3. A piece of knowledge is analogous to a word
    4. Defining the spirit world
    5. Formation of spiritual entities
    6. All spiritual entities are thought-like
    7. The spirit world is essentially thinking our physical world into existence
    8. Defining the spirit, soul, and spiritual expression
    9. Instincts and intuition, where do their messages come from
    10. The source of the spirit world's creative powers (from the spiritual model's standpoint)
    11. The spirit and soul will contain a record of the individual's life
    12. Defining the spirit and soul of nonliving things

13. The opening statement in Reference 1 describing what is in Reference 1 is as follows

4.2. The Workings in the Spirit World as Modeled by the Spiritual Model..126
    1. The source of the spirit world's creative powers (from the spirit world's standpoint)
    2. Knowledge is among the most powerful things that exist
    3. The spirit world creates everything
    4. Things that spiritual entities enabled to exist
    5. The larger the spiritual entity the more complex would be its translated form that could exist or be expressed in our physical world
    6. The number spiritual entities would increase something like two times an exponential rate of increase with every new piece of knowledge generated and added to the spirit world
    7. The spirit world could enable trillions upon trillions of things to exist in our physical world
    8. Continuously growing and changing long-living things need periodic adjustments:
    9. Each new universe will be different from any that came before it
    10. Multiple simultaneous universes and/or sequences of universes might be needed to restore balance
    11. We humans do not create things, not even artists. But what we are able do could still be called our creativity
    12. We could imagine or think about something only if that something is already created by the spirit world
    13. We form our thoughts, decisions, attitudes, purposes, etc. in the spirit world with our spirit and we carry them out in our physical world with our body
    14. The size of the spirit and the size of the brain of a living thing are not necessarily related to each other
    15. Why do we humans need sleep, why do we dream, and why are dreams usually surrealistic?

16. We are able to find and grab thoughts quickly enough to carry on a conversation
17. We could do a variety of things because our spiritual senses function extremely fast. This even includes letting the spirit world do a bit of our thinking for us
18. The text of a book must already exist in the spirit world before an author could compose it
19. Oneness pervades the spirit world
20. The spirit world is physically nowhere in our physical world, but it is spiritually everywhere in our physical world
21. The spirit world is in a sense more compact than is our physical world
22. Only one copy of anything that is not a "building material" would exist in our physical world
23. Nothing is perfect, not even the spirit world

4.3. A Natural Connection and Numerous Opposing Attributes Exist Between the Spirit World and Our Physical World........157

Chapter Five: Spiritual Qualities that likely Initiated the Spirit World and Gave Life to Living Things..........................161

5.1. Spiritual Qualities with Spiritual Forms that Are Not Part of the Spirit World..161
5.2. The 4Qs Likely Initiated the Existence of Knowledge and the Existence of the Spirit World................................164
5.3. Because the Spirit World Is a Living Thing, Almost Everything It Creates Is a Living Thing; Some Being More Living than Others..166
5.4. The 4Qs Initiated the Existence of the Spirit World as a Living Thing for a Reason..168
5.5. The 4Qs Most Likely Initiated the Existence of the Spirit World Multiple Times..169
5.6. The Size and Complexity of a Spiritual Entity Do Not Matter When It Comes to Being a Living Thing or Not..........170
5.7. A Possible Source of Why a Male Component and a Female

	Component Are Needed to Produce Living Things	170
5.8.	Living Things Are Most Certainly to Exist on Other Hospitable Planets Besides on Earth	171
5.9.	A Spiritual Place Exists Beyond the Spirit World That Is the Origin of Life	171

Chapter Six: The Three Major Obstacles That Are a Part of Our Physical World 173

6.1.	Our Physical World Was Designed to Be Opposite the Spirit World	173
6.2.	Spirituality and Science Need to Be Pursued Together, not Separately	174

Chapter Seven: How to Fulfill Mankind's Primary Purpose for Being Here on Earth 179

7.1.	The Spirit World Needs to Relearn How to Restore Balance Each Time an Imbalance Develops	179
7.2.	A Certain Requirement Must Be Met for the Restoration of Balance to Work	181
7.3.	What Does Mankind Need in Order to Fulfill Its Primary Purpose	183
7.4.	How Mankind's Secondary Purpose Will Be Fulfilled	184
7.5.	A Manager of a Complex World Needs Help to Balanced its State of Knowledge	185
7.6.	What Could Be Involved in Restoring Balance Besides Mankind's Primary Purpose	189
7.7.	Generating New Pieces of Knowledge that Restore Balance that Are Also Compatible with How Life Is in the Spirit World	192
7.8.	Of Particular Interest Is Why Evil in General Exists on Earth	194
7.9.	Past and Future Universes Are More Likely to Be Physical Instead of Nonphysical	197

Part Two

Implications of the Spirit World and the Spiritual Model Presented in Reference 1

Chapter Eight: Gradations and Diversity Are Among the Most Important Attributes of the Spirit World..........................201

Chapter Nine: Mankind's Gradations and Diversity Constitute a Gift from the Spirit World..........................205

- 9.1. Variations Are Sources of Strength and Abilities for Living Things..........................205
- 9.2. The Spirit of Each Living Thing in Any Universe Is a Portion of the Spirit World and Thus, the Living Thing Could Do Certain Things the Spirit World Could Do..........................208
- 9.3. Mixed-Breed Living Things Are Medically Stronger than Thorough Bred Living Things..........................209
- 9.4. Could and Should a Super Human Be Produced Through Genetic Engineering..........................211
- 9.5. Mankind's Gradations and Diversity Constitute a Gift to Mankind from the Spirit World..........................213
- 9.6. Gradations and Diversity Could Always Enable Mankind to Come Up with a Superior Version of Just about Anything..........................215
- 9.7. Taking Science into Consideration to Formulate More Complete Spiritual Models..........................217

Chapter Ten: Anything Reasonably Formulated Would Be Correct but Incomplete..........................221

- 10.1. Incompleteness Is a Natural Attribute of Everything..........................221
- 10.2. The Driving Force Behind Why Mankind Keeps Doing Research..........................223

10.3.	Better to Say "It Is Correct but Incomplete" Than to Say "It Is Wrong"	226
10.4.	Saying Something Is Either Right or Wrong Is More Expedient for Some Situations, Even if Less Accurate	227
10.5.	Would Things Unreasonably Formulated Be Incorrect	229
10.6.	Formulations Could Be Correct but Inappropriately Incomplete	231
10.7.	Dealing with Gradations as Gradations vs. Dealing with Absolutes	232
10.8.	The Spiritual Model Presented in Reference 1 Is Correct but Incomplete	235
10.9.	The Model of Everything Is Not Possible to Be Formulated by Humans	236
10.10.	A Crazy Quilt Made of Models and Has Lots of Holes	237
10.11.	Statements Could Be Interpreted in Different Ways Because They Are All Incomplete Even if Correct	239

Chapter Eleven: Evolution Spans All Past Universes; Living Things Emerge by Retracing 241

11.1. The Spirit World Periodically Brings New Universes into Being and Uses Retracing to Have Living Things Emerge in a New Universe 241
   1. The level of sophistication, creativity, and cleverness in the design of our universe and of the living things that reside in it
   2. Relative extent of evolution among living things on Earth
   3. Evidence of retracing
   4. Retracing is concept separate from conventional notions about evolution
   5. A lack of fossils of missing links could also indicate numerous past universes have come and gone and that evolution progressions span over past universes

11.2. Two Kinds of Evolutionary Progressions Exist 248
   1. Evolutionary progressions of the first kind
   2. Evolutionary progressions of the second kind

        3. How evolutionary progressions of the first kind work
        4. How evolutionary progressions of the second kind work
11.3.   The Process of Retracing..255
11.4.   Incompatibilities Could Form Among Living Things.............258
11.5.   Humans Are Likely to Have Been Evolving Only
        Moderately Long..259

Chapter Twelve: An Analogy Between Evolution and Maturation 263

12.1.   Maturation..263
12.2.   An Analogy Exists Between Our Species Evolutionary
        Process and Our Personal Maturation Process....................264
12.3.   Learning from Our Maturation Process and Shaping
        Our Species Evolutionary Path...264

Chapter Thirteen: Hang-Ups...271

13.1.   General...271
13.2.   Hang-ups Could Severely Influence a Person's Behavior......271
13.3.   Hang-ups Could Disrupt Mankind's Efforts to Fulfill
        Its Primary Purpose..273
13.4.   We Don't Have to Get Rid of Our Possessions......................275
13.5.   What Kind of Person Would Be Good to Work for................281

Chapter Fourteen: Our Choice of Dominant Basis for How We Carry Out Our Lives..283

14.1.   Ego and the Desire to Enhance Survivability.........................283
14.2.   Empathy and Compassion for One Another..........................285

Chapter Fifteen: Combining Spirituality and Science in Our Pursuit of Advancements of Both Kinds..........................287

Chapter Sixteen: More Discoveries and Developments the Spiritual Model Could Explain...291

16.1. Several Discoveries and Developments Were Presented in Earlier Chapters..291
16.2. The Reincarnation of Spiritual Entities that Serve as Spirits for Living Things...291
16.3. Possible Coincidence that Could Be Perceived as a Sort of Reincarnation of a Spirit..293
16.4. The Coincidence in the Preceding Section Could Be Explored with Clones...293
16.5. Spiritual Commonalities among "Sixth Sense" Phenomena the Spiritual Model Could Explain....................294
    1. Telepathy, intuition, instincts, cross-learning, dreams, channeling, and fortune telling
    2. Seeing and reading auras, remote viewing, what certain animals could see, and seeing or sensing ghosts
    3. Out-of-body experience, near death experience, pseudo space travel, pseudo universe travel, and space and universe travel in general
    4. Dreams, restoration process, long-term memory, and short-term memory
16.6. Matter and Anti-Matter..311
16.7. Artificial Intelligence..312

Chapter Seventeen: The Spirits and Minds of Young People Are Our Greatest Treasure for They Are the Future of Mankind (To Help Young People Develop Critical Thinking Skills).............................315

17.1. Helping Young People Prepare for Their Roles in the Future...315
17.2. What Are Critical Thinking Skills..316

Chapter Seventeen addresses the importance of developing critical thinking skills in children. This is because the future of mankind is in the hands of today's children. Therefore, we need to do what we can to help them avoid making the mess that members of earlier generations have made.

Part Three

Things that Could Become Commonplace in the Future

Chapter Eighteen: Studies and Explorations for Getting More in Touch with Our Spiritual Senses.................321

18.1. Studies and Explorations to Help Us Learn How to Get More in Touch with Our Spiritual Senses.........................321
18.2. It Might Not Be Difficult to Get More in Touch with Our Spiritual Senses..................326
18.3. Interactions and Interplays Between Advancements and Applications Will Accelerate Advancements....................327

Chapter Nineteen: Things We Could Sense in the Spirit World.....331

19.1. Things that Reside in the Spirit World........................331
19.2. The Origins of Things that Exist in Our Physical World........332

Chapter Twenty: Things We Are Likely Able to Do by Going Through the Spirit World..........................335

Chapter Twenty One: Will Jesus Come Again?..................339

21.1. Mankind Has Not Made a Concerted Effort to Evolve Further.........................339
21.2. The Problem Is Mankind Got Trapped by the Three Major Obstacles That Are a Natural Part of Our Physical World........341
21.3. What Does "Being Next to Jesus" Really Means....................342

**Chapter Twenty Two: Mankind's Potential Future
Way of Life in Partnership with the Spirit World**..............................345

22.1. What Mankind Is Doing in the Present..................................345
22.2. What Mankind Could Gradually Be Doing in the Future.......348
22.3. Mankind Was Likely Quite In-Touch with the Spirit
World but in a Limited Way, During Mankind's
Early Existence on Earth..352

Preface

The first reason this book is written: This book is in a sense a continuation of Reference 1 *"Connections with the Spirit World"* in that it presents new discoveries and developments that are made with the spiritual model presented in Reference 1 after Reference 1 was published.

The new discoveries and developments are quite numerous. Therefore, to help identify them more easily, they are summarized in the special section that immediately follows this Preface.

Presented in Reference 1 is a spiritual model that is unlike any other spiritual model that exists at the time. Spirituality was revisited in a manner never done before by taking into consideration engineering logic. The result is a spiritual model that is capable of providing reasonable and explicit explanation for just about any common everyday observations and experiences we could think of. No other spiritual model is able to do this in the manner this spiritual model could.

Thus, in this sense this spiritual model is able to model life more completely and more accurately than any other spiritual model has been able to do. This should not be a surprise since by taking into consideration engineering logic, the formulation of the spiritual model was working with a more complete set of attributes than if only spiritual attributes were considered.

Science related concepts and spirituality are usually handled separately such that the study of one usually does not take into consideration the other. The spiritual model presented in Reference 1 is the first spiritual model that took into consideration a science related concept in its formulation. The result is a spiritual model that is capable of explaining things such as; why we need sleep, why dreams tend to be surrealistic, how telepathic communication works, how some individuals could see and read auras, why we have instincts and intuition, etc.

The second reason this book is written: Feedback from readers of Reference 1 indicates it was difficult to read because it is packed full of unconventional concepts such that it reads more like a textbook. Therefore, the second reason this book is written is to present some of the major points in Reference 1 in an easier to read format. Therefore, in a sense this book could serve as an introduction for Reference 1 even though it is a continuation of Reference 1. Accordingly, a reader might want to read this book before reading Reference 1.

The third reason this book is written: The third reason for writing this book is because the author is concerned about the future of mankind in view of mankind's current overall poor behavior. Mankind has been wasting a lot of time, energy, and resources waging wars and battles of all kinds, hating and mistreating among one another, and carrying out bad applications of its technical advancements. Not only is mankind negatively spending time, energy, and resources by such poor overall behavior, it also spending a lot of time, energy, and resources having to deal with the complications and the mess that are caused by such poor overall behavior.

According to the spiritual model, our physical world was created by the spirit world for a purpose. This means mankind has a primary purpose for being here on Earth, and the primary purpose couldn't possibly be to do the bad things mentioned. Therefore, mankind is not at all fulfilling its primary purpose. There has to be a way to inspire and motivate mankind to behave overall in a manner that would lead fulfilling its primary purpose.

This book presents a way to do this, and it is different from what has been tried and hasn't worked. What has been tried repeatedly essentially consists of rules imposed upon mankind by mankind. Presented in this book is a way to inspire and motivate mankind from within mankind to want to behave overall better. It does this by providing mankind with a more complete understanding about life.

Up to now, mankind has only half of the knowledge it needs about life. As with anything we do, if we have only half the knowledge needed to know how to do something, we are likely to do it poorly. That is the situation mankind has been in; i.e., presently, mankind has mainly the physical part of the knowledge about life and very little of the spiritual part.

This book presents a way to help mankind gain the rest of the knowledge about life, which is the spiritual part of the knowledge about life. By having more completely the knowledge about life, mankind would more likely be inspired and motivated from within itself to behavior overall better.

Mankind's primary purpose: This book explains mankind's primary purpose for being here on Earth. Mankind's primary purpose is to help the spirit world restore balance in its state of knowledge. Our physical world was created such that this purpose would be fulfilled if mankind were to do the following:

Mankind carries out its life simultaneously in the spirit world and our physical world. Thus, mankind is to learn as much as possible about the two worlds and about how things work in each of them. Then mankind is to apply all this knowledge positively and constructively to make mankind's life on Earth to be as close as possible as how life is in the spirit world. Life in the spirit world is indicated by the spiritual model in Reference 1 to be naturally filled with love, empathy, and compassion. This is because of the oneness that naturally pervades the spirit world as explained in detail in Reference 1 and also summarized in this book.

New Discoveries & Developments Made with the Spiritual Model Presented in Reference 1

Numerous discoveries and developments made with the spiritual model presented in Reference 1 are presented in Reference 1. Additional discoveries and developments were made after Reference 1 was published and are presented as follows.

1. The Source of Life

The four spiritual qualities (4Qs) that reside outside the spirit world likely exist before knowledge and the spirit world exist. The 4Qs very possibly initiated the existence of knowledge and thus also the existence of the spirit world. The 4Qs are also the qualities living things would have, and therefore they are the "source of life" for living things. Chapters One, Five, and Nineteen.

2. Mankind's Purpose for Being Here on Earth

Mankind's purpose for being here on Earth has two parts:

The primary purpose is to help the spirit world restore balance in its state of knowledge.

The secondary purpose is to generate as many new pieces of knowledge as possible to help the spirit world continue to grow.

Discoveries and developments lead to the following summary statement regarding how mankind is to go about fulfilling its primary purpose for being here on Earth

Mankind's primary purpose for being here on Earth is to gain as much knowledge as possible about our physical world and the spirit world, and to apply that knowledge to make mankind's life on Earth to resemble as closely as possible to how life is in the spirit world.

Fulfilling the secondary purpose is essentially automatic. Chapters One, Two, Four, Six, Seven, and Twelve.

3. The Brain Communicates with the Spirit

Functional MRI studies of brain activity patterns by Dr. Marcel Just of the Carnegie Mellon University at Pittsburgh in my opinion support the notion that the brain communicates with the spirit. This gives additional confirmation for the spiritual model presented in Reference 1. Chapter One.

4. Additional Confirmations of the Spiritual Model

Additional confirmations of the spiritual model presented in Reference 1. Chapters One, Nine, and Sixteen.

5. Our Physical World Is Opposite the Spirit World

Many of the attributes of our physical world are opposite those of the spirit world. The spirit world designed our physical world to be that way so that our physical world could help restore balance in the spirit world. Chapters Two, Four, Six, Seven, and Twenty One.

6. A Democratic Governing System Is Fragile

A democratic governing system is fragile because mankind's current human nature is such that we humans have two desires that are incompatible with each other.

On one hand, we humans want to have the freedom that a democratic governing system provides.

On the other hand, we humans are not particularly interested in carrying out the required responsibilities that come with the freedom a democratic governing system provides.

These incompatible desires are human behaviors that are a part of mankind's current state of human nature. The result is a democratic governing system would not receive the complete support it needs. Consequently, a democratic governing system is fragile not because it is inherently fragile, but it is because of mankind's current state of human nature. Chapters Two and Three.

The concepts described in this item appear to be supported by Luke Kemp's work as presented in the next item.

7. Our Civilization Might Be at Risk of Ending

Luke Kemp of the University of Cambridge studied the lifespan of 87 ancient civilizations covering 3000 BC to 600 AD, a span of 3,600 years, and found the average lifespan was 341 years ranging from 24 years to 1100 years. The common reason for each civilization to end was human behavior; i.e., degradation in the support of the civilization. Chapter Two.

The alarming thing is the civilization of the United States of America is 244 years old in year 2020, and support of our civilization has definitely degraded. Over the time span from year 2016 through year 2020, various members in government holding powerful positions are breaking numerous laws and rules of our governing system and are getting away with it. This is mainly because approximately one third of the population don't seem to care and even seem to support such bad behavior. Chapter Two.

8. Gradations and Diversity Constitute a Gift to Mankind

Gradations and diversity, and thus variations, constitute a gift from the spirit world to mankind.   Chapter Nine.

Variations are sources of strength and abilities for living things.

Gradation and diversity could always enable mankind to come up with a superior version of just about anything.

Explained in terms of the spiritual model presented in Reference 1 as to why inbreeding could produce off-springs with medical associated weaknesses.

9. Two Kinds of Evolutionary Progressions Exist

Two kinds of evolutionary progressions exist. Chapter Eleven.

Experiences are constantly being gone through by some living thing either within the spirit world or in a universe. Thus, new pieces of knowledge are constantly being generated and added to the spirit world. Therefore, the spirit world is constantly becoming more advanced. This means any new living thing it designs at any point in time could be made more advanced than any new living thing it designed in the past. This is denoted in Chapter Eleven as evolutionary progressions of the first kind. How this works in terms of the spiritual model presented in Reference 1 is described in Chapter Eleven.

Evolutionary progressions of the second kind would happen to living things as they reside in a universe. I.e., it would be what we would perceive as being the usual evolutionary progressions. How this works in terms of the spiritual model presented in Reference 1 is also described in Chapter Eleven.

10. An Analogy Between Evolution and Maturation

An analogy exists between Evolution and maturation, and we could use this analogy to help us understand how we could control how we are evolving. Chapter Twelve.

11. Studies and Exploration to Get More in Touch with Spiritual Senses

Various studies and explorations are considered in terms of how we could learn how to get more in touch with our spiritual senses. It might be easier than we think. Chapter Eighteen.

12. An Alternative Interpretation of Being Next to Jesus

An alternative interpretation of what it means to "being next to Jesus" and what the concept of "Jesus returning again" could mean. The traditional interpretation is in terms of physical situations whereas the alternative interpretation is in terms of spiritual situations. Chapter Twenty One.

In terms of spiritual situations, 'being next to Jesus" would mean we have done the good things that would make the makeup of our spirit to be similar to that of the spirit of Jesus. Thus, our spirit would automatically spiritually go to being next to the spirit of Jesus within the spirit world.

In terms of spiritual situations, "Jesus returning again" would mean by our making the makeup of our spirit to be similar to that of the spirit of Jesus, our spirit would be the one that spiritually go to the spirit of Jesus within the spirit world instead of Jesus physically coming to us in our physical world. And, as indicated, our spirit would automatically spiritually go to the spirit of Jesus within the spirit world.

These alternative interpretations would be more compatible with today's world in which we humans are now more into science and technology than in the past when concepts such as the two discussed were first formulated.

Special Acknowledgement and Dedication

This book is dedicated to my parents Chester and May. It is often said we are not likely to appreciate all the things our parents have done and gone through until we are parents ourselves. I could go further by adding "- - - and we have grown old ourselves".

Our generation lives in essentially a different world than they did. We had our struggles as well, but ours are so different from theirs that we did not fully appreciate their dreams and hopes, their emotional and psychological strengths, their persistency and dedication to excellence that they had against all odds, and what they had to do to succeed. My father was orphaned at age thirteen and inherited a sizeable debt from his parents. He worked the next thirteen years to pay off the debt and to save up enough to then marry the girl who would become our mother.

They started out virtually penniless in this country, had little education in their native country, were unable to speak English, and faced discrimination. They made a great team even though they were ten years apart in age. They were extremely honorable, creative, intuitive, and resourceful. They found ways to make things work in spite of a very limited income, and they managed to bring up five well behaved children

They never complained about their struggles or considered themselves as having had a harder time than anyone else. Looking back, it is amazing how they did it, and whether we realize it or not they have been a major source of inspiration and motivation for all five of us.

Acknowledgements

My greatest love and appreciation go to my parents Chester and May for finding a way after they have pass away to tell my sibling and me that an afterlife exists. They made the four strange events described in Chapter Two of Reference 1 to happen shortly after my mother pass away three years after my father. They clearly indicated that an afterlife exists. This was a gift more precious than anything physical.

My most precious love goes to my deceased wife Mae for her patience and understanding as I spent time working on Reference 1 and for continuing to be with me spiritually after she passed away. She continues to be with me more than eleven years later as I struggled with my own cancer and after effects. By her continuing to do this she is providing me with additional evidence that an afterlife exists.

I express my deepest gratitude for the love and support I receive from my widowed oldest sister Mabel, my older brother Walter and his wife Eleanor, and my youngest older sister Pauline and her husband Bob. They have helped to keep me going through the years. Also helping to keep me going is my very best friend since childhood Joe, a very thoughtful and caring person.

Because the world today is changing so fast, my children and grandchildren have become an important source of learning for me. I very much value this in addition to the love and support they give me. Much appreciation goes to my very thoughtful, talented, and creative son Michael, my extremely able and clear thinking daughter Catherine, her super capable and versatile husband Jon who holds everything together in their family as Catherine needs to go on frequent business trips, and their intelligent, energetic, and talented teenaged sons Jake and Zack. The ever-present support of my niece Denise Hom Berry, late husband

Dave, and their daughter Jackie has helped my late wife and me a lot and is very much treasured.

I very much appreciate the help R. Ken Coit has given me that enabled me to find time to work on Reference 1 and this book. He has helped countless other individuals in his caring approach to life and with his charitable activities. I very much enjoyed the stimulating and inspirational conversations regarding spirituality with Eric Johnson and the Rev. Tommy E. Smith Jr., the author of Reference 10. I particular appreciate the roles Prof. Robert F. Steidel and Prof. Karl S. Pister have played in my life.

My warmest and very special thanks go to my very special friend Rachel Burke who was instrumental in helping me understand the meanings behind the four strange events presented in Chapter Two of Reference 1. I regret she and I lost contact with each other after I retired and I was unable to locate her again. I hope Reference 1 and this book will somehow find their way to her.

A very appreciative thanks to Eric, Peggy, and Alex Johnson for their wonderful work putting this book together. This especially applies to Alex who designed the text layout and developed the very creative design for the front and back covers.

Part One

The Spirit World, the Spiritual Model, and Mankind's Primary Purpose on Earth

Chapter One Background Information and Examples of New Discoveries and Developments

Chapter Two: Our Spiritual Advancements Need to Match Our Technical Advancements

Chapter Three: Mankind's Current State of Human Nature

Chapter Four: The Spirit World and the Spiritual Model Presented in Reference 1

Chapter Five: Spiritual Qualities that likely Initiated the Spirit World and Gave Life to Living Things

Chapter Six: The Three Major Obstacles That Are a Part of Our Physical World

Chapter Seven: How to Fulfill Mankind's Primary Purpose for Being Here on Earth

Chapter One

Background Information and Examples of New Discoveries & Developments

1.1. Introduction

In the book "Connections with the Spirit World" by this author, Reference 1, spirituality was revisited in a manner never done before by taking into consideration engineering logic in the process. The result was the formulation of a spiritual model presented in Reference 1 that is unlike any other spiritual model that existed at the time. The spiritual model is capable of explaining things other spiritual models were unable to explain, and it could do it in manner that is more logical and explicit than any other spiritual model could.

Up to now, spirituality and science have always been treated separately; i.e., one was assumed to not have anything to do with the other. The spiritual model presented in Reference 1 is the first to include engineering logic in its formulation. The model achieved a high level of confirmation by being capable of explaining in a logical and explicit manner just about any common everyday observations and experience we could think of.

For example, it could explain:

1. Why we need sleep.

2. Why dreams tend to be surrealistic.

3. How telepathic communication works.

4. How some individuals could see and read auras.

5. Why we have instincts and intuition.

6. Our consciousness, intelligence, and mental abilities reside with our spirit and not with our brain as is usually assumed.

7. How it is we would have an afterlife after we die.

8. Etc.

Reference 1 contains numerous discoveries and developments that were made during the formulation of the spiritual model in addition to ones listed above. The primary reason for writing this book is to present the new discoveries and developments that were made with the spiritual model after Reference 1 was published.

Feedback from readers indicates Reference 1 was difficult to read because it is packed full of unconventional concepts, and thus it reads more like a textbook. Therefore, a second purpose for writing this book is to present some of the major points in Reference 1 in a manner that is easier to read. In this sense this book could serve as an introduction to Reference 1. Therefore, a reader might want to read this book before reading Reference 1, even though this book is in a sense a continuation of Reference 1.

A summary of some of the major points in Reference 1 is presented in Chapter Four of this book. To help make Chapter Four easier to follow, Chapter Two and Chapter Three are placed ahead of Chapter Four to introduce some of the concepts that appear in the summary.

A third purpose for writing this book is to present spiritual reasons that could hopefully inspire and motivate mankind to improve its overall behavior, and stop doing things such as starting wars and battles of all kinds, threatening and mistreating one another, doing bad applications of technical advancements, committing criminal acts of all kinds, etc.

The spiritual model indicates our universe was designed and brought into being to fulfill two purposes. This means we humans, being the most intelligent living things on Earth, have the following two purposes to fulfill for being here on Earth, as pointed out in Reference 1.

1. **Our primary purpose is to help restore balance in the state of knowledge of the spirit world. Explained in this book are the following:**

    a. Why maintaining balance in the spirit world's state of knowledge is vitally important.

    b. How to go about fulfilling this primary purpose.

    c. Fulfilling this primary purpose will require a high level of intelligence, creativity, and focus. The reason the spirit world selected humans to fulfill this primary purpose is because we are capable of doing it, but we need to be focused and stay focused. So far, we have not been focused on fulfilling this primary purpose. Therefore, something has to be done to get us going.

2. **Our secondary purpose is to generate as many new pieces of knowledge as possible to help the spirit world continue to grow.**

    a. Fulfilling this secondary purpose will almost be automatic. Nevertheless, it is good to point out we humans have this secondary purpose.

    b. The reason this will almost be automatic is because we humans are curious about things and therefore like to explore and experiment such that we would be going through experiences almost constantly. Experiences would automatically generate new pieces of knowledge.

1.2. The Source that Gives Life to Living Things

The spirit world is where the spirit of a living or nonliving thing is created and resides. We might ask, do nonliving things have spirits too? According to the spiritual model presented in Reference 1 they do, and this is contrary to conventional thinking. Every living and nonliving thing that exists in our physical world exists because something in the spirit world enables it to exist. This something is the spirit of the living or nonliving thing.

Regarding a living thing that exists in our physical world, the following applies:

1. The spirit world creates a spirit that is a living thing in the spirit world, and it would be the spirit of a living thing that resides in our physical world. This spirit is what enables the living thing to exist in our physical world. It begins at the moment of conception of the living thing.

2. Conception takes place simultaneously in the spirit world and in our physical world. The life of the living thing begins in both the spirit world and our physical world. It continues thereafter with one part being carried out in the spirit world and the other part being simultaneously carried out in our physical world.

3. The spirit makes its decisions and forms its thoughts, intentions, etc. in the spirit world, and it then "pilots" the physical body in our physical world to carry out the decisions, thoughts, intentions, etc. In this sense, the spirit of the living thing is what the living thing really is.

4. The concept of the spirit "piloting" the physical body is explained in Chapter Four and in more detail in Reference 1. Briefly, it goes as follows:

    a. When we drive a car, we are analogous to the spirit and the car is analogous to our body.

b. When we become proficient enough at driving, the car could feel as if it is a part of us, as if we and the car are "one", even though we are who we are and we are not we plus the car.

c. We are designed such that our spirit is always piloting our body as long as the body is alive. Thus, during the first couple years of life, our body would feel as if it is a part of our spirit, as if our spirit and our physical body are "one," even though our spirit is really who we are.

d. We are really who we are in the sense that our consciousness, intelligence, mental abilities, memories, emotions, etc. reside with our spirit and not with our brain and body as usually assumed. This is explained in Chapter Four and in more detail in Reference 1.

5. When the part of our life in our physical world is over, the part in the spirit world would continue in the spirit world thereafter forever. This part is often referred to as being the afterlife, although it is simply the continuation of the part that has been going on in the spirit world.

A similar sequence of statements could be made regarding a nonliving thing. However, there is an extra twist.

Some spirits that the spirit world creates would be "perceived as being" nonliving things, and some would be "actually" nonliving things. A spirit perceived as being a nonliving thing would enable what we would perceive as a nonliving thing to exist in our physical world such as a chair, table, pair of shoes, etc. Spirits that are actually nonliving things are not likely capable of enabling anything to exist in our physical world except for the basic materials and energies that make up our physical world.

Thus, the materials and energies that make up our physical world and things that we would perceive as nonliving would have a part residing in the spirit world as well as a part existing in our physical world just as living things would. The difference is the spirit of any of such things would not be making decisions or forming thoughts, intentions, etc. in the spirit world and would not be piloting the physical thing that resides in our physical world.

The reason I say "perceive as being" a nonliving thing is because according to the spiritual model presented in Reference 1, everything, aside from the materials and energies that make up our physical world, that exists in our physical world would have some level of life, as explained in Chapter Five of this book and in Reference 1. Some things are more alive than others, and the lowest level of life is "elemental life." For example, this would be the level of life a chair, table, pair of shoes, etc. would have.

An example of actions something that has elemental life could do would be how water could be perceived as being conscious of its temperature and would be perceived as knowing when to freeze, melt, or evaporate. A similar thing could be said about how various chemicals are conscious of the condition they are in and thus know when to react and when to not react. More examples are given in Chapter Five and in Reference 1.

A living thing's body is made up of a whole lot of different things that have elemental life. Each kind of such things would have its own kind of elemental life. When all the elemental lives of the things making up the body are acting together, they would make up the life of the living thing.

Vital organs would function, the heart would pump, eyes would see, muscles would work, etc. It is for this reason that actions of things such as water and chemicals could be perceived as having elemental life.

What is described here is likely to apply in some manner in other universes as well.

We might ask; what makes one spirit a living thing in the spirit world and another a nonliving thing in the spirit world? The answer is there is a "source of life" that is a separate thing from the spirit world, and it gives life to living things. This is a major new discovery and development that was made with the spiritual model presented in Reference 1 after Reference 1 was published.

Since this chapter is an introduction for this book, this new discovery and development is mentioned only briefly here. More details are presented in Chapter Five.

The main point is that the initiation of life is a spiritual phenomenon and not a physical phenomenon. Life cannot be initiated by physical means alone such as certain chemicals coming together and then being hit by lightning, as speculated at one time. Scientists recreated such situations in a laboratory, and they found the process doesn't initiate life.

A major attribute of this new discovery and development, as explained in Chapter Five, is that this thing we are calling the source of life most likely exists before knowledge and the spirit world exist. Therefore, it is likely what initiated the existence of knowledge and thus also the existence of the spirit world as a living thing.

Details regarding how this is likely the case are presented in Chapter Five. Also explained is why calling this thing the "source of life" seems to made sense even before the idea came up that it is likely what initiated the existence of knowledge and the spirit world.

This could answer the lingering question as to what it is that initiated life on Earth and most likely also on other planets and perhaps also on some moons in our universe.

For the rest of the discussions in this book, we will simply refer to something that has only elemental life as being a nonliving thing.

1.3. The Earliest Form of Recognizable Life on Earth Has to Be the Simplest Possible Living Thing

The earliest form of recognizable life on Earth has to be the simplest possible such as slimes, as generally speculated.

Every new universe is unique and different as explained in Chapter Four such that the spirit world would never have worked with the new materials and energies making up the new universe. Thus, the spirit world would need to learn how to work with the new materials and energies before it could do much with them. Therefore, the spirit world would begin by doing the simplest things first and then gradually do more complex things as it gradually increases its knowledge about what could be done with the new materials and energies.

This could explain why the early living things on Earth were very primitive and why living things gradually become more complex and advanced as time goes by and as evolution progresses. The gradual nature of the process means the spirit world is going through a learning process. This is as indicated by the spiritual model presented in Reference 1.

1.4. Our Universe Is Huge and Is Filled with All Sizes of Galaxies, Stars, Planets, and Black Holes for a Reason

Before the spirit world was even ready to initiate recognizable life on Earth, it has a lot of learning and preparatory work to do. Our universe began as a cloud of basic building materials and energies such as subatomic particles, electrons, protons, and neutrons. The spirit world needs to learn how to form all the atomic elements as necessary out of this cloud in order to be able to start making up living and nonliving things.

This could explain why our universe has almost an infinite number of galaxies, stars, planets, and black holes and why they come in all sizes. The spirit world is bound to need to do a lot of exploring and experimenting with all kinds of ways to make atomic elements. The spirit world probably discovered that it needs a variety of sizes of stars and black holes in order to make all the necessary atomic elements it needs to construct living and nonliving things.

Therefore, the preparatory work the spirit world most likely had to do would include the following:

1. Countless explorations and experiments the spirit world needed to carry out to find out how to make all the atomic elements that are possible to make.

2. Discovering the need for diversity in the sizes of galaxies, stars, planets, and black holes in order to produce all the atomic elements that are possible to be produced.

    More specifically, the spirit world would form essentially an infinite number of stars ranging from the minimum size possible to the maximum size possible in order to produce all the kinds of elemental atoms possible to produce and in the amounts of each that are needed.

    a. The smaller stars would produce the lighter elemental atoms.

    b. The largest stars would produce the heavier elemental atoms.

    c. All the in between size stars would produce all the in between size elemental atoms.

3. Deciding to have multiple other sites in our universe beside Earth to generate the kinds of new pieces of knowledge the spirit world needs to restore balance in its state of knowledge. This helps to assure at least one site would fulfill the purpose of our universe.

4. Such sites are likely purposely located very far apart such that

one site would not influence another in a manner that could cause none to fulfill the purpose of our universe.

5. The large separations would also assure independent thinking and the development of a diversity of possible ways to prepare our physical universe for housing living things.

6. This means while we are looking and listening for possible life to exist somewhere in our universe, we are not likely to receive such signals until the highly intelligent life has ended at such sites and on Earth. I say this because:

    a. When the spirit world is ready to initiate highly intelligent life to exist in our universe it is likely to do so more or less at the same time at every such site.

    b. The sites are purposely located billions of light-years apart for the reasons stated in Item 4 and Item 5.

    c. Therefore, by the time any signals generated by highly intelligent living things residing on any of the other sites reach Earth, those living things on that site and mankind on Earth would likely no longer exist.

7. An exception might apply to those sites that are making good progress in fulfilling the need of the spirit world.

    a. The highly intelligent living things at such sites would not have wasted their time, energies, and resources making wars and other conflicts among each other as we humans have done and are still doing on Earth.

    Therefore, they are able to develop practical ways to do space travel without the use of fossil fuels.

    b. They could be riding in the UFOs and UAPs that have been

visiting Earth.

c. They would be wise enough not to interact with us humans but to only observe to see how we are doing, and realize we are not doing very well.

1.5. Functional MRI Studies of Brain Activity Patterns Support the Notion that the Brain Communicates with the Spirit

For the past decade Dr. Marcel Just, Reference 2, of the Carnegie Mellon University at Pittsburgh has been using functional MRI imaging to study the brain activity patterns formed when people think about specific things. For individuals with normal functioning brains he found the patterns are the same for every person when they are thinking about the same thing such as a screw driver. Normal functioning brains would be those of individuals who are not, for example, autistic.

Communication between brain and spirit was not the topic of Dr. Just's studies. However, in my opinion, his findings support the notion that the brain of an individual communicates with the spirit of that individual as predicted by the spiritual model presented in Reference 1. The following would be how such communications work, according to the spiritual model.

1. The certain thing that an individual is thinking about would have a spiritual form that is a certain spiritual entity in the spirit world.

2. Each piece of knowledge that it is a part of the spiritual entity would be issuing its unique signal. The combination of all the pieces of knowledge making up the spiritual entity would be unique. Therefore, the signal issued by that spiritual entity would be unique.

3. When an individual is thinking about a certain thing, his or her

spirit would use its spiritual senses to find the spiritual entity in the spirit world that is the spiritual form of that certain thing. His or her spirit would also sense the signal of that spiritual entity with its spiritual senses.

4. The spirit would then send that signal to the individual's brain. Upon receiving that signal the brain would form a certain activity pattern.

5. Dr. Just had several individuals think about the same object. Every one of those individuals formed the same brain activity pattern. This means their spirits were all sensing the same signal from the same spiritual entity and were sending the same signal to their brains, and thus their brains all formed the same activity pattern

This indicates that the brain and the spirit communicate with each other as predicted by the spiritual model presented in Reference 1.

This also further confirms the validity of the spiritual model even though providing such confirmation was not the purpose of Dr. Just's study.

1.6. The Way Things Work in the Spirit World Is Very Positive for All Religions and All Perceptions of God

A very positive thing could be brought about by the spiritual model presented in Reference 1 being able to logically and explicitly indicating that the spirit world exists, that it is extreme large, and that it encompasses all religions, all denominations of religions, and all perceptions of God.

Up to now because there are so many different religions, so many denominations of religions, and so many perceptions of God that the whole situation is confusing as to which is correct. In addition, the way religions are being conveyed to the population is becoming increasingly removed from today's increasingly high-tech way of life. The continuing decline in church attendance suggests that the stories and messages provided by the various churches are losing their inspirational appeal, especially since young people are becoming increasingly educated and into high technologies.

The spiritual model presented in Reference 1 could help revitalize mankind's interest in religions. The spiritual model brings spirituality closer to how life is in today's increasingly high-tech world by being able to do the following:

1. The spiritual model brings spirituality closer to how life is in today's increasingly high-tech way of life by taking into consideration engineering logic in its formulation and by coming up with a model that could do a whole lot more in explaining common everyday experiences and observations, including high-tech things, than any current spiritual model could.

   The spiritual model is able to do this in a logical and explicit manner. This should make its messages more appealing for young people, and thus some of them might become more interested in spirituality and religions.

   Spirituality and religion have been treated as being separate from science such that they have not been taken into consideration in scientific pursuits. I tend to agree religion should not be considered in science, but I think spirituality should be.

   I say this because the spirit world brought our universe into being and it included science as being a part of our universe. For this reason, science and spirituality naturally go together.

2. The spiritual model clearly indicates in a logical and explicit way that the spirit world is real. It does this by indicating something somewhere somehow has to exist that knows how to enable our

universe and everything in it to exist.

This notion and the concept that the spirit world is made of all existing pieces of knowledge is likely to appeal with today's young people who are increasingly into high technology.

3. The spiritual model indicates that the spirit world encompasses all religions, all denominations, and all perceptions of God. It is very possible for something like the spirit world to be larger than all of the religions and all perception of God combined and thus could encompass all of them.

This means all religions, all denominations, and all perceptions of God are equally valid since each is a portion of the spirit world, and the spirit world is real and valid.

Therefore, ideally, this could promote constructive interactions among the various religions. For example, the messages of one religion could augment the messages of another and this could be beneficial for both religions. Similar positive and constructive interactions could take place among all religions, all denominations, and all perceptions of God.

I think religions were first formulated to improve mankind's overall behavior, and they have been successful to some extent. However, as our way of life drastically changed, their effectiveness gradually declined because their stories and messages are gradually losing their inspirational appeal in today's fast-moving high-tech way of life.

For this reason, I think all religions need to be updated. The update should focus on inspiring and motivating mankind from within itself (instead of imposing rules from outside of itself as usually done) to improve its overall behavior based on notions and concepts that are relevant to today's way of life.

1.7. Brief Statement Regarding How to Fulfill Mankind's Primary Purpose for Being Here on Earth

The spiritual model presented in Reference 1 explains the following regarding the spirit world:

1. What the spirit world is made of and how things work in it.

2. How it is that the spirit world is a growing and learning thing and thus is a living thing.

3. How it is that a natural partnership is formed between the spirit world and each of us, and we are to make this partnership a mutually benefiting thing.

4. It thus follows that our universe was designed and brought into being by the spirit world for a purpose, and that mankind is here on Earth to help our universe fulfill its purpose

5. Therefore, mankind has a primary purpose for being here on Earth.

How to fulfill mankind's primary purpose is deduced from the spiritual model presented in Reference 1. The deduction process is presented in Chapter Seven of this book. A quick statement of mankind's primary purpose is as follows. A more detailed discussion about this is presented in Chapter Seven.

Mankind's primary purpose for being here on Earth is to help restore reasonable balance in the spirit world's state of knowledge. How to go about fulfilling this is as follows:

1. **We are to learn everything we can about our physical world and how things work in it.**

2. **We are also to learn how to work around various obstacles that**

are a natural part of our physical world. We are not to get trapped by these obstacles.

3. Shortly after we start learning about our physical world, we are to also start learning everything we can about the spirit world and how things work in it.

4. We are to apply all the knowledge we gain from doing Items 1, 2, and 3 to accomplish one main thing. It is to make mankind's life on Earth to be as close as possible to being how life is in the spirit world.

5. All the experiences mankind would have to go through to achieve this would automatically generate the kind of new pieces of knowledge that would help restore balance in the spirit world's state of knowledge. More details on how this works is presented in Chapter Seven.

As explained by the spiritual model presented in Reference 1, because of what the spirit world is make of and the way things work in it, life in the spirit world is naturally filled with love, empathy, and compassion.

Such qualities are simply a natural attribute of the spirit world. This is because, according to the spiritual model presented in Reference 1, living things are all directly or indirectly a part of one another in the spirit world. By contrast, living things are all separate entities in our physical world.

Therefore, making mankind's life on Earth to be as close as possible to being like how life in the spirit world is not going to be easy. This means to do this mankind would have to seriously use all of the following in order to make mankind's life on Earth to be as close as possible to being how life is in the spirit world.

1. All of its mental abilities.

2. All of its physical abilities.

3. Its imagination

4. Its creativity.

5. Its ability to generate as much as possible all of the mentioned kinds of knowledge.

6. Its ability to apply at all times in positive and constructive ways all the knowledge it has generated.

We humans are here on Earth to in essence do all of this learning for the spirit world. Our universe was designed such that all the new pieces of knowledge we humans would generated by going through the experiences of making mankind's life on Earth to be as close as possible to being how life is in the spirit world would automatically help restore balance in the state of knowledge of the spirit world.

Each human spirit is a portion of the spirit world. This means what each of us learn is being learned by that portion of the spirit world and thus becomes a part of the spirit world.

Therefore, this means whatever the spirit world becomes, each of us has a part in making it happen for better or for worse. Accordingly, all of us need to take care of the spirit world by doing what is good and avoid doing what is bad. Otherwise, the spirit world could become increasingly unbalanced, unable to form wisdom, and unable to keep itself viable. Thus, it would eventually vanish, and everything it created would also vanish, including us humans.

This is explained in detail in Reference 1 and is also outlined in Chapter Four of this book.

Our universe was designed such that the kinds of experiences we need to go through in order to do all this learning would be available for us to go through and thus do this learning for the spirit world. These particular experiences happen to need a physical universe in order to be formed. They could not be formed with a nonphysical universe or in the spirit world, that itself is nonphysical. This is why the spirit world had to design and bring into being our particular physical universe and why our physical universe has the particular attributes it has.

As the presentation continues, we will see how it is that the spirit world would periodically need to design and bring into being a new universe and why every new universe would naturally be different from any that were designed and brought into being in the past. Most new universes are likely to be physical and some would be nonphysical. Each would be different in a variety of ways such as being dimensionally different, made of different kinds of building materials, have different forms of energies, be uniform throughout instead of having separate galaxies and stars, etc. It all depends on what the spirit world needs at that moment in order to restore balance in its state of knowledge.

1.8. How Far Along Is Mankind in Fulfilling its Primary Purpose

We might ask; how far along are we humans in fulfilling our primary purpose, since it is obvious that we have not yet fulfilled it. The answer is that we have a long way to go. We did not learn how to work around the obstacles that are a natural part of our physical world. Instead we got trapped by them and thus the spiritual part of our mental evolutionary process got stalled.

Thus, we got so engrossed with battles, conflicts, power and status seeking, etc. the spiritual part of our mental evolution slowed down to a snail pace, if it is even advancing at all. Meanwhile the technical part of our mental evolution surged ahead. But without an adequate amount the spiritual part of our mental evolution to go along, we humans do an enormous amount of bad application of our technical advancements. This made our battles, conflicts, power and status seeking, etc. even worse, and our lives increasingly complicated.

Consequently, instead of trying to get ourselves out of the trap, the way we humans have been behaving for thousands of years has kept us trapped by the obstacles that are a natural part of our physical world, and thus the spiritual part of our mental evolutionary process remain stalled. It is time for us humans to do something to get mankind out of the trap and to restart the spiritual part of its mental evolutionary process such that we humans could resume becoming the much better species we are likely meant to become.

The spiritual model presented in Reference 1 provides a way for us to begin doing this. Reference 1 also presented numerous discoveries and developments associated with the spiritual model that could further help us, and this book presents additional discoveries and developments that are made after Reference 1 was published.

A major part of the presentations in the ten chapters following Chapter Seven is directly or indirectly about the additional discoveries and developments associated with the spiritual model made after Reference 1 was published. These discoveries and developments reinforce what is briefly stated early in this chapter and is stated in detail in Chapter Seven as being mankind's primary purpose for being here on Earth.

However, since mankind has been behaving overall poorly for thousands of years, behaving overall poorly has become almost a permanent part of its current state of human nature. Therefore, mankind needs to seriously do something to improve its state of human nature as a part of restarting the spiritual part of its mental evolution.

This means mankind needs to put in a sustained and significant effort to improve mankind's overall behavior in addition to putting a significant effort to fulfill mankind's primary purpose.

In addition, no matter how much knowledge our pursuit of spiritual advancements could give us regarding how to behave overall better, we still need to make a serious and sustained choice to behave overall better. Otherwise we would keep falling back into the trap repeatedly and the spiritual part of our mental evolutionary process would keep getting stalled repeatedly.

The terms "technical advancement" and "spiritual advancement" were introduced in Reference 1 and are also used in this book. Technical advancement is knowledge gained about how things work in our physical world, and spiritual advancement is knowledge gained about how things work in the spirit world. These advancements are the topics of the next chapter.

Chapter Two

Our Spiritual Advancements Need to Match Our Technical Advancements

2.1. It Is Time We Improve Mankind's Overall Behavior

Have we ever wondered how it is that we humans are able to achieve amazing technical advancements but are unable to achieve much improvement in our overall human behavior, which remains generally poor for thousands of years? We have major wars, long lasting lesser battles, conflicts over the same kinds of issues repeatedly, various groups mistreating and/or hating various other groups, and criminal acts of all kinds.

If we are intelligent enough to achieve technical advancements we should be intelligent enough to avoid wars and battles, resolve conflicts peacefully, address issues in a wise and reasonable manner, stop doing criminal acts, and realize it is not good for the future of mankind for various groups to mistreat and/or hate various other groups. But so far, we have not been able to behave in such better ways consistently or in a lasting manner.

How we got into this situation is understandable and can be explained. Also, can be explained is we could have easily done better. In our past, as in our present, we always have numerous evolutionary paths available for us to choose to go on at any time such as the following:

1. Some paths would lead to our pursuing technical advancements much more than spiritual advancements, as we have been doing

up to now. We would then be confronted with numerous serious major issues and problems that come with insufficient spiritual advancements to go with our massive amount of technical advancements, as has been happening to us for thousands of years.

2. Some paths would lead to our pursuing spiritual advancements much more than technical advancements, and relatively minor issues and problems would come up such as inconveniences, a lack of more technically advanced methods to resolve various complex problems, and a lack of more technically advanced ways to treat various medical and other conditions.

3. Some paths would lead to our pursuing technical advancements and spiritual advancements more or less comparably, and we would have far fewer and far less pressing issues and problems than those of option 1, and perhaps somewhat fewer and some what less pressing than those of Option 2. In other words, Option 3 would be the most balanced and therefore the best of the three options.

We are choosing among such paths almost every moment whether we are aware of it or not, and we are frequently switching from one path to another as our situations and conditions change. Sometimes we do it with contemplation, but mostly we do it without much thought because we are busy with other demands. We are likely to assume life will always be the way it is, and we would likely automatically choose mostly paths of the Option 1 kind. Consequently, mankind's spiritual advancements remain lacking and thus our overall behavior remains consistently poor. This then leads to mankind making a mess on Earth, particularly regarding human, national, and international relations. Based on the spiritual model presented in Reference 1, we could have avoided much or all of the mess if we had chosen mostly paths of the Option 2 or Option 3 kinds.

We might ask the following two questions:

1. Why would mankind's overall behavior be generally poor and make a mess on Earth because we choose mostly evolutionary paths of the Option 1 kind?

2. What exactly are spiritual advancements, how come we have not achieved much of them, how could we go about achieving more of them, and why would we even want to achieve more of them?

We sort of know or suspect we have a spirit and a soul. However, their roles in our life are vague mainly because they are vague as to what they are. But, one thing seems clear; i.e., if we have an afterlife it must be our spirit and our soul that are living it since our body, and particularly our brain, would be deceased. Evidence is presented in Reference 1 to indicate an afterlife does exist for each of us. Therefore, each of us must have a spirit and soul.

The spiritual model in Reference 1 is able to provide a plausible explain what a spirit is and what a soul is, and the model is also able to explain what their functions are in our life and what their functions are in our afterlife.

Some individuals, including myself, have experienced events that indicate afterlife definitely exist. I had four such events as described in Chapter Two of Reference 1. The events indicate that in our afterlife we would still have our consciousness, thinking ability, reasoning ability, memories, etc. This means all such mental abilities must reside with our spirit and not with our brain since our brain would be deceased. This is contrary to common beliefs.

The spiritual model in Reference 1 is able to explain in a logical and explicit manner how all this works. This spiritual model is different from all other spiritual models that currently exist. This is because it takes into consideration engineering logic and reasoning in its formulation instead of relying strictly on faith for acceptance as is the case for all the other spiritual models.

The spiritual model in Reference 1 thus explains that while we are alive in our physical world, we are carrying out our life simultaneously in the spirit world where our spirit resides and in our physical world where our body resides. According to the model, our spirit forms our thoughts, decisions, and intentions in the spirit world, and it pilots our body to carry them out in our physical world.

In order to form the wisdom necessary to do a good job of carrying out life while we are alive in our physical world, we need to be knowledgeable about both parts of our life. Right now, we know a lot about the physical part but not much about the spiritual part. This imbalance prevents us from being able to form the necessary wisdom regarding how to carry out our life. Consequently, we humans as a whole would often form our thoughts, decisions, and intentions without a whole lot of wisdom. Our overall behavior reflects this and it is also indicated by the mess we made on Earth.

No matter the task we face, if we lack sufficient and reasonably balanced knowledge about how to perform it, we are not likely to perform it very well or wisely. The knowledge would be reasonably balanced if it could lead to an understanding as to what to do and what not to do, what is good and what is bad, what is fair and what is unfair, etc. Notice how the instructions that always come with a complex piece of equipment would include what not to do as well as what to do in order to help assure safe, wise, and proper use and maintenance of the equipment.

We gained a lot more knowledge about our physical world than about the spirit world. This is understandable since our consciousness is tuned much more to our physical world than to the spirit world, and our five major senses function in our physical world but not in the spirit world.

According to the spiritual model in Reference 1, we also have an undetermined number of spiritual senses that function in the spirit world, but we are not tuned to them as much as we are tuned to our five major senses. Therefore, we do not know exactly what they are and how many we have, even though we use them constantly to find thoughts, solutions, concepts, etc. in the spirit world. We are tuned to them enough to find such things but we are not tuned to them enough to do much else. How we use our spiritual senses to find things in the spirit world is explained in greater detail in Reference 1.

Consequently, it is easy for us humans to develop a significant imbalance in our knowledge about life. A significant imbalance did happen and thus we are not able to form the wisdom necessary to do a good job of carrying out our life. This explains why we humans could make the mess we made on Earth. To reduce this imbalance, we need put more effort into gaining knowledge about the spirit world. We need to be equally knowledgeable about both the spirit world and our physical world.

What is holding us back is the fact that we are unable to sense what the spirit world is made of and how things work in it, whereas we are able to sense what our physical world is made of and how things work in it. For this reason, it is much easier to learn a lot about our physical world, and it is much harder to learn just about anything about the spirit world.

I think the afore mentioned condition is purposely designed to initially exist with us so that we could survive our early existence when we lived in caves. During this period, we had to be extremely conscious and knowledgeable about the physical dangers that surrounded us, and we had to be able to come up with ways to protect ourselves. Being able to focus on the physical part of our life and to be able to pursue technical advancements were vital during this early period.

In addition, I think part of the reason we are created with our high level of intelligence is so that we could eventually figure out how to develop ways to determine what the spirit world is made of and how things work in it, and thus be able to pursue spiritual advancements as well as technical advancements. We could then start increasing our knowledgeable about the spirit world so as to decrease and eventually eliminate the imbalance in our knowledge about life.

But instead, we got trapped by some obstacles that are a natural part of our universe. We were supposed to learn how to work around the obstacles and not get trapped by them. The obstacles are among the numerous features that have to be a part of our universe in order for our universe to serve the purpose of helping to restore balance in the spirit world. Had we been doing what we should have been doing to work around them we would be helping our universe fulfill its purpose for having been brought into being. This and the three major obstacles that are a part of our universe are explained in greater detail later in this book and also in Reference 1.

The spiritual model presented in Reference 1 provides a way to determine what the spirit world is made of and how things work in it. It achieved a high level of confirmation by being able to provide plausible spiritual explanations for every common everyday experiences and observations we can think of. Thus, it enables us to gain knowledge about the spirit world in a manner similar to how we gain knowledge about our physical world. We could then become more able to form the wisdom necessary to do a good job of carrying out our life and also improve our overall behavior. Therefore, we could gradually stop making a mess on Earth and also gradually clean up the mess we have made.

2.2. A New and More Effective Way to Pursue Spiritual Advancements

Our current spiritual models, such as the various religions, traditions, customs, ways of life, etc. enable us to achieve a certain amount of spiritual advancements. But it has not been enough to inspire and motivate us humans enough to behave overall a whole lot better. We need a more effective way to pursue spiritual advancements. Preferably, the amount of spiritual advancements achieved should be comparable with the amount of our technical advancements achieved so that our knowledge of both our physical world and the spirit world would be more reasonably balanced. This would be more consistent with how it is that we are carrying out our life simultaneously in both worlds.

The spiritual model presented in Reference 1 provides a new and more logical and explicit way of pursuing spiritual advancements. We would be able to understand more clearly what we are doing as we are pursuing spiritual advancements. This is because the spiritual model indicates what the spirit world is made of and how things work in it. Current spiritual models tend to be vague in this respect, and thus they pretty much rely strictly on faith for their acceptance. It thus follows that the spiritual model in Reference 1 offers an approach to pursuing spiritual advancements that is similar to how we have been pursuing technical advancements.

In my mind, the spiritual model presented in Reference 1 provides the first step in this new approach to pursuing spiritual advancements. Since this approach is similar to how we pursue technical advancements, as we proceed, we are bound to find ourselves able to formulate increasingly more advanced spiritual models similar to how we are able to formulate increasingly more advanced technical models as we pursue technical advancements. Each more advanced spiritual model we are able to formulate would represent an increase in our understanding of the workings of the spirit world just as each more advanced technical model we have been able to formulate represents an increased understanding of the workings of our physical world.

Sometimes a more thorough examination of an existing spiritual model or technical model could reveal something we did not recognize earlier. This is how in many cases new technical discoveries are made by further applications and/or examination of an existing technical model. This happens in just about every technical field, most notably in the high technology field, medical field, astronomical field, materials technology field, etc.

It even happened in what we might consider as the spirituality field. More specifically, a further examination of the spiritual model presented in Reference 1 revealed what must be accomplished in order to fulfill mankind's primary purpose for being here on Earth. This was logically and explicitly deduced from the spiritual model. How to fulfill mankind's primary purpose is briefly presented in Chapter One as an opener to this book, and it is described in greater detail in Chapter Seven.

As far as I know, this is the very first time we have been able to figure out exactly and explicitly what must be accomplished in order to fulfill mankind's primary purpose is for being here on Earth. The explanation given in Chapter Seven seems to make a lot of sense, and it is consistent with what I think every sensible person in the world would wish everyone on Earth would eventually be like someday.

If a variety of progressions are possible with our pursuit of technical advancements, they are bound to be also possible with our pursuit of spiritual advancements. After all, our physical world was designed and brought into being by the spirit world. Therefore, if we find our physical world amazing, the spirit world is bound to be more amazing in order be able to design our physical world to be as amazing as it is.

Therefore, a variety of progressions can be expected to occur as we proceed with our pursuit of spiritual advancements. Each progression in either pursuit serves as a kind of "platform" from which to launch to achieve the next progression. The next progression could come in the form of new discoveries with the application of an existing model or it could come in the form of a new and more advanced model. This is how it has been with our pursuit of technical advancements, and I would expect it to be how it will be with our pursuit of spiritual advancements.

2.3. Interactions and Interplays Between the Spiritual Parts and the Technical Parts of Things

Up to now we have been talking about the pursuit of spiritual advancements and the pursuit of technical advancements as separate activities. Spirituality has been commonly perceived as being separate from science. Thus, scientists have essentially excluded spiritual considerations from their explorations. I think by doing this we are missing out on a lot of discoveries that we could otherwise make.

A prime example is presented in Reference 1 in which by taking into account spiritual considerations as provided by the spiritual model in Reference 1, quantum superposition and entanglement could be explained as to how they could happen whereas up to now they have not been able to be explained with science alone. The explanation presented in Reference 1 thus revealed that the phenomena are partly spiritual and partly technical.

Life is partly spiritual and partly technical since we are carrying out our life simultaneously in the spirit world and our physical world. Also, our physical world exists because the spirit world designed and brought it into existence. Accordingly, there is a natural connection between the spirit world and our physical world. Therefore, when mankind focuses mainly on the part of life that is in our physical world while ignoring or not being aware of the part of life that is in the spirit world, mankind is bound to miss out on a whole lot about life. And, in my opinion this is exactly what is happening and this is one of the reasons why we made a mess on Earth. The spiritual part of our mental evolutionary process has gotten stalled and is still stalled.

In most cases we are dealing with things and issues that are either largely spiritual with a negligible amount of technical or largely technical with a negligible amount of spiritual. Therefore, we have been able to get by without taking into account together both the spiritual part and the technical parts of what we are doing. But, our "getting by" does not mean we are "doing well." There will be cases in which neither part would be negligible and by not taking into account both parts together we might "get by" for a while but in the long run we would not "do well."

For example, our continued technical development of automation has a significant spiritual part that seems to be largely ignored by the developers. A collision is in the making among the continued elimination of jobs due to automation, the creation of new jobs that are fewer than the ones eliminated, the new jobs requiring a lot more college education, the decreasing affordability of college education, and the continuing increase in the world population and thus the need for more jobs. When we take into consideration all such factors, we might find automation to be advantageous technically but a bad idea spiritually. Therefore, to assure a good future for mankind we might want to be selective about what to automate and how far to go with automation.

Another example, the development of social media initially seems spiritually wonderful because it could enable everyone on Earth to be more in touch with one another. But there is a technical consideration that seemed to have been ignored by its developers, or they might just did not think about it. The ignored technical consideration consists of all the possible bad ways social media could be applied. There seems no limit to the possible bad ways social media could be applied. They have become a serious problem, and the things we must do to prevent being a victim of bad applications have significantly complicated life and eroded our trust in the high technology ways of doing anything. The net result is it is not clear whether social media is a positive thing or a negative thing.

If we ignore either the spiritual part or the technical part for everything we do, then we are doing so even when neither of the two parts is negligible. We would then be shutting out any interactions and interplays between spirituality and science that happened to be important, and thus any wisdom that could come from such interactions and interplays would be missed. This would add to the mess we have already made.

2.4. Differences Between the Part of Life in the Spirit World and the Part of Life in Our Physical World

As discussed earlier, we are simultaneously carrying out our life partly in the spirit world and partly in our physical world. Major differences exist between these two parts of our life. A better understanding of the differences would help us understand how the overall behavior of mankind could lack spiritual qualities. It would also help us realize mankind has been too focused on the physical part of life and has not been paying enough attention to the spiritual part of life. Hopefully, this information would inspire and motivate mankind to start being more serious about pursuing spiritual advancements.

Accordingly, as we continue to pursue spiritual advancements, we will discover things about the spirit world that are bound to be more interesting than what we discover about our physical world. For example, according to the spiritual model in Reference 1, things we now perceive as paranormal are bound to be normal and natural in the spirit world. What we are able do with the mechanisms behind such things is likely to be more fascinating than anything we have been able to do with the mechanisms behind the things we find in our physical world.

Because everything in the spirit world is directly or indirectly a part of everything else, how things work in the spirit world would be very different from how things work in our physical world. For example, activities that go on in the spirit world are more likely to be long-term long-range instead of short-term short-range whereas the opposite is more common in our physical world. This is because things are separate and more isolated in our physical world.

The spirit world is also nonphysical and dimensionally unlimited. Communication is instantaneous, and it goes on throughout the entire spirit world. New spiritual entities are constantly being formed; thus, there is no end to new discoveries that could be found in the spirit world. All this and more are indicated to be the case according to the spiritual model in Reference 1.

Two major differences between the spirit world and our physical world are particularly worth pointing out. They could have us rethink what life is about.

1. **Our physical bodies are separate entities in our physical world, while our spirits are all a part of one another in the spirit world:**

    a. We might want to rethink whether we should live our life on Earth as separate entities as we are doing or as a part of one another as we are in the spirit world.

    - Since both are true, it is best to live our life as both. In other words, while we are physically separate, we can be spiritually a part of one another.

    - This means being kind, empathetic, and compassionate with one another.

    - This also means valuing differences. As explained in Chapter Nine, diversity is a gift to us from the spirit world.

    b. Because every human spirit shares a part of itself with every other human spirit, every human body is constructed internally and externally in accordance with the same basic plan.

    c. Also, all humans have basically the same kinds of feelings, emotions, consciousness, and major senses.

    d. Such sharing in the spirit world extends also to the spirits of animals and various other living things on Earth.

    e. This is why the bodies of animals and various other living things are constructed similarly to how human bodies are constructed. They also have the same feelings, emotions, consciousness, etc. as humans do, but only in different intensities; i.e., some are more intense and some are less intense.

    f. This could explain why humans have more empathy and compassion about animals than we would about, for example, fishes, birds, and insects. Our spirits have more in common with the spirits of animals than with the spirits of fish, birds, and insects.

2. **Our physical life lasts only for a while, but our spiritual life lasts forever:**

   a. While being separate entities would last only for a while, being a part of one another would last forever.

   b. Therefore, instead of living our life for the part of our life that is in our physical world where it lasts only for a while, it would be wiser to live our life for the part of our life that is in the spirit world where it lasts forever.

   c. In fact, according to the spiritual model in Reference 1, we need to live our life for the part of our life that is in the spirit world at least to some degree in order to be able to figure out what it is we are to do to fulfill our primary purpose for being here on Earth.

   d. Accordingly, we might want to make our most important achievements in life to be things that mean something in the spirit world and not to be things that mean something only in our physical world.

   e. When we are near the last days of our life, we are more likely to agree with this. When individuals who are close to death were asked what they wished they have done in their life that they did not do. The most common response was they wished they have been kinder to other people.

All that has been discussed so far could explain why it is that most of those who had a near-death experience would describe the place they were in as being filled with feelings of love, empathy, and compassion and that everything about anything is perfectly clear while they were there. They often wanted to stay there instead of returning to our physical world. Their description of that place matches how the spirit world would be like according to the spiritual model presented in Reference 1.

According to the spiritual model, the following are attributes of the spirit world:

1. The spirit world at any point in time is made of every piece of knowledge that exists at that point in time.

2. Each piece of knowledge has a connection of the first kind with every other piece of knowledge that exists at that point in time. There are also connections of the second kind, and these would make up the soul of a living or nonliving thing. This is explained in Chapter Four.

3. Only one copy of every existing piece of knowledge is kept in the spirit world.

    a. Any duplicates would immediately merge and only one copy will remain.

    b. Because everything in the spirit world is made of pieces of knowledge and only one copy of every piece of knowledge is kept, everything in the spirit world is naturally directly or indirectly a part of everything else in the spirit world.

4. As a result of everything being directly or indirectly a part of everything else, the feelings of love, empathy, and compassion naturally pervade the spirit world.

5. Because every piece of knowledge that exists at any point in time resides in the spirit world, everything that we are able to know about anything that exists in the spirit world at that point in time would be perfectly clear at that point in time.

a. This is because all the knowledge that exists about anything at that point in time is available there.

b. This explains why individuals who had a near death experience would say that everything was perfectly clear in the place that they were at.

6. The descriptions by individuals, who had a near death experience, of the place where they have been during their near-death experience support what is said in Items 1 through 5.

Science is commonly perceived as being separate from spirituality. But again, according to the spiritual model in Reference 1, everything making up our physical world is brought into being by the spirit world. Science is a part of our physical world. Therefore, science has to be a part of spirituality, and by keeping science and spirituality separate we would be dealing with both in an incomplete manner.

This could explain why the various existing ways of handling spirituality, such as with the various religions, have not been very effective in inspiring and motivation us humans to behave overall better. It is because spirituality has been handled in an incomplete and therefore not an overwhelmingly convincing manner.

The spiritual model in Reference 1 has the potential of getting us to rethink how we are handling spirituality and how we could make our currently existing spiritual models, such as the various religions, more complete and thus more convincing and more effective in getting us humans to behave overall better.

Conversely, our not taking into consideration spirituality in our scientific pursuits could be why we have not been able to satisfactorily explain certain things about our physical world. As presented earlier, the previous inability to explain how quantum superposition and entanglement could happen are examples of this. Also as presented earlier, we have not been able to determine what dark matter and dark energy are and what causes gravity to exist. These unanswered questions might also be the result of leaving spirituality out of our scientific explorations.

Further discussion about the spiritual model in Reference 1 is presented in Chapter Four. For more details about the spiritual model and its formulation process please see Reference 1, in particular the second edition in which the typos in the first edition are corrected and a "Preview/Index" is added to help clarify many of the unconventional notions and concepts that are part of the formulation process for the spiritual model.

The main text of Reference 1 takes up slightly under 500 pages, and the "Preview/Index" takes up slightly over 100 pages. In order to present the model and to show how it is able to achieve a high level of confirmation by being able to capture reality to a significant degree, and because the formulation is highly unconventional and thus needs extensive explanation, the main text ended up taking up almost 500 pages.

Spirituality was revisited in Reference 1 in a manner never done before by taking into consideration engineering logic and reasoning. The result is a spiritual model that is unlike any other that exists up to now. It is able to provide spiritual explanations in a logical and explicit manner for any common everyday experiences and observations we can think of.

2.5. A Civilization at Risk

A concern that many of us in the United States of America have in years 2019 and 2020 is that we could be facing a constitutional crisis. Several powerful members in our government are breaking the laws and rules set forth in our governing system, and the majority members of the Senate are supporting their political party more than they are supporting our governing system. These members are getting away with what they are doing. What is worse is approximately one third of the population doesn't seem to mind, and some seem to even like what is happening.

Chapter 2

The long-term long-range consequences are that such violations of laws and rules would encourage future members in our government to do the same. Thus, the meaning and power of our Constitution will degrade and our democratic way of governing will become more like a dictatorship. It is troubling that a large segment of the population seems unable to understand this as they seem to be living life for the moment without being concerned about the future of our nation.

What is worse is that the president is acting in a manner that seems to be purposely dividing the population rather than bringing the people together as a president should. It is hard to believe that dividing the population would serve what he says he wants to do and that is to "make America great again." The population ought to ask, could a divided nation ever be really great?

On the other hand, based on the work of Luke Kemp in Reference 3 what is happening might be a natural thing that happens to nations in general based on the histories of past civilizations.

Numerous nations in the past have come and gone, and there appears to be a commonality in the reasons they have ended, and it has to do with the behavior of mankind. This is one of the reasons Reference 1 and this book is written. It is an attempt at inspiring and motivating mankind to improve its overall behavior. Hopefully such improvement could happen before our civilization comes to its end, because our civilization is starting to show some signs that it could be starting to end, as described as the presentation continues.

Luke Kemp of the University of Cambridge studied the lifespan of 87 ancient civilizations covering 3000 BC to 600 AD, a span of 3,600 years, Reference 3. The average lifespan for the 87 civilizations was 341 years, ranging from 14 years to 1150 years. In 2020 the age of our civilization is 244 years starting with the establishment of our Declaration of Independence on July 4, 1776. Several factors are listed by Kemp as possible contributing factors for the ancient civilizations to end.

In my opinion the most important of the factors he listed is that the attitude of the citizens has gradually changed to where they no longer support their governing system enough to sustain it. Obviously, any decline in the citizens' support of their governing system is a bad sign for the survivability of a civilization. Therefore, it is disturbing to see indications that this is beginning to happen with our governing system in years 2017 through 2020. For example:

1. The checks and balance responsibility each of our three branches of government has as a means to keep all three branches funtioning as coequals is being ignored by numerous individuals in powerful positions in our government. These individuals are thus disrupting the way our governing system is supposed to function. Approximately one third of the population is supporting such behavior of these individuals, and this makes matters even worse.

2. A majority of the Senate is behaving as if it were a part of the executive branch instead of as part of an independent and coequal branch of our governing system as the Senate is supposed to be.

3. The attorney general is acting as if he is a personal attorney for the president instead of as the attorney general of the people as his position is supposed to be.

4. Certain powerful members in our governing system are breaking the laws and rules of our governing system and members of their political party don't seem to care. They are more concerned about keeping their jobs than they are of supporting our governing system.

5. Our governing system is now 243 years old. The current generation of the population is more or less ten generations removed from the founding generation of our governing system. We can tell, most of them are not very knowledgeable about the how our governing system is setup to function. Consequently, our

governing system is not being supported by the population as it should be.

6. For example, a surprisingly and disturbingly large portion of the population doesn't seem to mind that the laws and rules of our governing system are being violated. They might not be knowledgeable about the laws and rules, but they still don't seem to care even when the laws and rules are pointed out to them. It is as if they like to see things being "blown up" in any manner even if it is against the law. Thus, it seems when they have someone in a very powerful position "blowing things up" in our governing system they will support that person. The following two reasons might be why.

7. A large portion of the population apparently felt disenfranchised and thus voted for candidates who would shake up our governing system.

    a. For one thing the income and wealth disparity are enormous and continue to worsen.

    As pointed out by Bernie Sanders, Reference 4, the three wealthiest individuals have a combined wealth equal to the combined wealth of half of the population consisting of the least wealthy in our nation. Something is wrong with this.

    b. The homeless problem keeps growing and no one has come up with a workable solution. The enormous income and wealth disparity bound to have something to do with this.

    This portion of the population is thus happy to see various members of government not following laws and rules of our governing system. They are thus part of what is disrupting the way our governing system is supposed to function.

8. There might also be an even larger portion of the population that is simply frustrated about almost everything and they simply want to blow things up. Some indications of this includes the following:

a. Mass shootings are occurring somewhere almost once a week.

b. Teenage and young people suicides are increasing significantly.

c. Movies and video games featuring massive explosions are popular.

d. The high-tech advancements might make certain things work easier and faster but because of a lot of bad applications of such advancements they make life more complicated with more things to worry about and to constantly monitor and update.

Also, when high-tech things go wrong or are hacked to stop working it takes a lot of time and money to get them fixed. Therefore, whatever amount of time is saved is more than taken up by what it takes to keep things running, not to mention all the frustrations involved.

e. Artificial intelligence and automation are doing away with a lot of jobs. A typical young person entering the work force could expect to have to change their field of occupation three times and thus needs to get reeducated and/or retrained each time over their lifetime.

f. Cyber-criminal behavior is as common as the air around us, and a lot of it is targeted at disrupting major institutions that are vital to our way of life. We tend to not hear about them because we have been pretty successful so far in stopping them, and I think the people in control do not want to alarm the population by revealing them.

9. Another major cause for general frustration is the exploding world population. It is getting so that someone or something is constantly getting in the way of whatever we are trying to do whether it is a traffic jam, long wait at the DMV, or trying to get to talk with a live person on the phone instead of prerecorded options. Our natural resources, especially clean water and food sources, are stretched to their limits.

The net result of all of the above is that people now tend to think more in terms of living and surviving for the moment and forgetting about long-term long-range consequences. In some regions of the world conditions are much worse. For example, rain forests are being destroyed to make room for growing food.

What Kemp said in essence is that every one of the 28 civilizations failed most commonly due to the behavior of its population plus some outside help such as **disease epidemic**, climate change, crop failure, famine, war, etc.

Talk about **disease epidemics**, in year 2020 the entire world was hit with the Covid-19 novel coronavirus pandemic that was so severe it virtually shut down the entire world. A pandemic is worse than an epidemic in that it covered the entire world whereas an epidemic is localized. To slow its spreading, people are to shelter in place at home, to go out only for essential needs, to stay at least six feet apart if they have to go out. Among all nations, the United States of America got hit the hardest. This adds to the concern that the United States civilization is at risk of ending in a manner similar to how past civilizations ended.

In my mind, Kemp's findings essentially mean every civilization will eventually and naturally fail mainly due to mankind's human nature. I would say mankind's "current state of human nature," because I believe mankind's human nature could evolve to be better or to be worse. I also believe mankind's current human nature would evolve to become better if we gain a good understanding of the spirit world; i.e., if we achieve a large enough amount of spiritual advancements and thus become more knowledgeable about life as a whole instead of mostly about the technical part of life.

In my mind, what Luke Kemp said implies that the attitude of the population gradually changed from initially working together to assure their civilization will succeed to later working for self interests. Thus, the civilization would eventually no longer receive the support necessary to remain viable. I suspect the population did not realize its attitude has changed, because the change was gradually happening from one generation to the next. Each generation loses more of the memory of the initiating generation. Several generations later, essentially all of the memory of the initiating generation is lost. It also doesn't help that we humans tend to think short-term short-range more than we think long-term long-range. The net result is the current population might not anticipate a possible end to their civilization until it is too late.

Therefore, I would imagine a possible way a civilization could end is as follows. However, every civilization is different so that the following is not intended to be a typical description of all the ones that ended.

1. In the beginning the personal needs of the population and the needs of the civilization to be viable would be mutually compatible.

2. The population is thus motivated to work together toward assuring their civilization would succeed. Differences in social and economic status would not matter much.

3. The generation that initiated the new civilization and the generation that immediately follows are likely to be very knowledgeable about the laws and rules of their governing system; e.g., they would be very familiar with the Constitution of their nation, assuming they have a Constitution as their document for their governing system.

4. The civilization eventually becomes well established and

reasonably wealthy largely because of hard work, dedication, innovations, and technical advancements. At the same time, the workings of their governing system have been "patched" countless times to resolve or work around problems resulting by bad behaviors of some people and numerous bad applications of technical advancements. Life has thus become complicated and messy. Disparities in social and economic statuses keep increasing and are creating increasing dissatisfactions.

5. The generation that now exists is many generations removed from the one that initiated the civilization. Thus, the current generation is not knowledgeable about the laws and rules set forth in the Constitution of their nation.

6. In addition, the population has grown to where it is more difficult to keep tract of all the law breakages that are committed. Therefore, the current population becomes used to seeing laws being broken that go unpunished, as if such behavior has become more the norm. For example, in the case of our civilization, traffic laws are being broken nonstop and enforcement is slim.

7. The net result of Items 5 and 6 is that the current generation doesn't seem to care if some members in their governing system are breaking the laws and rules of their governing system. Thus, their governing system is starting to not get the support it needs to stay viable.

8. The infrastructure is now old and in need of rebuilding and updating. But funds were not set aside for it largely because we humans tend to think short-term short-range more than long-term long-range.

9. Control of the governing system and corporations is increasingly held by the relatively few very rich while the rest of the population is struggling to make ends meet. Some even become

homeless. The population is thus unwilling to accept any increase in taxes and is increasingly bothered by the increasingly large social and economic disparities.

10. Meanwhile much of the civilization's wealth is "locked up" with the few very rich and not a whole lot is in circulation to maintain the strength and health of the civilization. Therefore, while the civilization is very rich and strong in terms of total dollar amount, it is actually much poorer and weaker because the amount of dollars in circulation is much less than the total dollar amount in the civilization.

A side note: This condition applies to our nation, the United States of America, in year 2020. The huge and growing disparity in wealth and income is becoming a significant problem. It contributes to the serious homelessness problem. It causes discontent as many in our population have to work multiple jobs to make ends meet. Again, as noted by Bernie Sanders, Reference 4, the three wealthiest individuals in our nation have a combine wealth that is equal to the combined wealth of half of the population consisting of the least wealthy in our nation.

11. A bad and growing homelessness problem has developed and is a clear evidence of declining health and a developing weakness in the civilization. The human energy, potential, and strength being dissipated due to homelessness are serious economically and psychologically even if they are not acknowledged as such by a large part of the rest of the population and by many in government.

Thus, it is not credible to say a civilization is rich and strong especially when the homelessness problem now includes whole families and college students as a sign that the civilization is not as rich and strong as it might seem. The problem is becoming increasingly complicated by the day. The effort to resolve the problem is far short of what is needed, and no effective long-term long-range plan is coming forth.

12. A large portion of the population has thus been pretty much left out of the prosperity enjoyed by the rest. The large economic disparity continues to grow. Marketable skills are changing faster than many in the population could get reeducated or retrained to acquire new skills. Opportunities for reeducation and retraining are not very accessible due to finances and/or locations. Thus, some simply can't find new employment, stopped looking for new employment, and are thus not counted as being among the unemployed, but they should be.

    Some who find employment are employed only part time. For a more accurate count of the unemployed, these individuals should be fractionally counted, but they are not. Some individuals need to work multiple jobs to make ends meet, and they too need to be somehow reflected in the unemployment count. A more accurate unemployment count would help formulate more effective actions to help those in need of help.

    A lot or all of individuals in such conditions are now seeking change, any change to find relief. So, they are willing to vote for any candidate who would shake things up when they get elected whether it is good for the sustainability of their civilization or not.

13. The measure of economic strength ought to take into consideration the conditions of individuals in such conditions, and it ought to also take into consideration the disparities in pay of the jobs people are holding instead of counting all jobs as being equal. An inaccurate indication of economic strength would naturally lead to governmental actions that do not match the needs of the civilization, and this could contribute to ending the civilization. Such a mismatch is serious but it does not seem to be recognized by all.

    The population in general and various members in government could feel the effects of this mismatch but they seem unable to understand the effect is caused by this mismatch. They are likely to feel helpless to fix the effects. Thus, their reaction is likely to be as follows.

14. The attitude of the general population and various members in government has now changed from supporting their governing system to going for self-interests. The civilization is now no longer receiving the support is used to receive.

15. The economy is meanwhile being artificially propped up short-term for the sake of self-interest of politicians so they can get reelected. They reduce taxes and interest rates while at the same time greatly increasing the national debt leaving the next generation with enormous interest payments and a significantly reduced amount left for crucial services and other needs. This kind of action is very destructive long-term, and the population generally doesn't seem to understand that it is, because people tend to not think long-term long-range very much. This too would contribute to ending a civilization.

16. Politicians can easily manipulate the population because of the population's tendency to not think long-term long-range, and some politicians would do just that. Even worse, the manipulation seems to usually work for a large portion of the population. This is probably because a large segment of the population is not aware of the dynamics of the economy and how it could be manipulated to appear to be good in the short run by essentially borrowing form from the future and thus creating a poorer future.

17. Members holding powerful positions in government are now behaving in a manner that would enrich themselves instead of taking care of the population and the civilization. It is as if they are grabbing what they could "before the ship goes down".

18. If all this continues, the civilization will eventually end largely due to the people's behavior; i.e., due to mankind's current state

of human nature. This would be as said by Luke Kemp regarding the 28 ancient civilizations.

If this scenario is more or less typical of what happens to civilizations in general, then mankind's current state of human nature is the primary cause and it needs to be improved. This means we need to achieve significantly more spiritual advancements than we have achieved so far, and we need start doing it pretty soon because indications are that our civilization might very well be at risk.

2.6. A Democratic Governing System Is Fragile and Needs Complete Support

This topic connects the end of this chapter, Chapter Two, to the beginning of the next chapter, Chapter Three. Chapter Two is about our spiritual advancements being lagging far behind our technical advancements, and Chapter Three is about the current state of human nature.

By the end of Chapter Three we will see how it is that these two subjects combined could result in mankind having two major desires that are incompatible with each other. In my opinion, this incompatibility is a major reason a democratic governing system is fragile. An equally important second reason is that we humans seem to be unaware of the first reason.

The two incompatible desires are as follows:

1. On one hand, we humans want to have the freedom that a democratic governing system provides.

2. On the other hand, we humans are not particularly interested in carrying out the required responsibilities that come with the freedom a democratic governing system provides. For example: a. Our voter turnout in elections is consistently low.

b. A typical individual does not take the time to be well informed about what is needed to know to make wise voting decisions.

c. We humans tend to think more short-term short-range than long-term long-range. This too is incompatible with a governing system that is meant to function long-term long-range and is meant to be for all of our nation's population and not just for any one person specifically.

d. Because of the current state of human nature, we have a tendency to see if we can break laws and rules and get away with it. A prime example is how we consistently break traffic laws and rules when we are driving, and we usually get away with it.

A recent example took place in April, 2020 when the world is dealing with the Covid-19 virus pandemic. Some young people of college age and younger were purposely breaking rules and to see if they could get away with it. It is also an example of how we humans tend to think more short-term short-range than long-term long-range. More details are presented in Chapter 3, Subsection 3.3. "Long-Term Long-Range Thinking vs. Short-Term Short-Range Thinking."

Such behaviors tend to get us into the mindset that we can break laws and rules in general and get away with it.

This also could explain why a large segment of the population currently doesn't seem to mind when certain members of government in powerful positions are breaking the laws and rules of our governing system.

Consequently, because mankind has these incompatible desires, a democratic governing system is fragile and can easily evolve into being a dictatorship when certain kinds of people get into powerful positions in government. This situation exists right now in our nation in years 2017 through 2020. Such an initial partial breakdown of our governing system could only continue to increase, and it would be very difficult to repair.

In my opinion, the "number two desire" almost invites certain kinds of individuals to get into a powerful government position to change a democratic governing system into a dictatorship one.

A real-life example of how a democratic governing system fairly quickly turned into a dictatorship form of government is touched upon in the documentary movie *"The Kingmaker,"* made in August 2019 and directed by Lauren Greenfield, a highly acclaimed documentary photographer and documentary filmmaker, Reference 5.

Chapter Three
Mankind's Current State of Human Nature

3.1. Mankind's Human Nature Is Changeable

We might ask: why I refer to mankind's human nature as mankind's "current state of human nature?" It is because any living thing's nature is changeable and could be shaped by its environment if the living thing is in it for an extended period. This is the case for creatures, and indications are that this is the case for us humans as well. The following are examples:

1. Creatures on the Galapagos Islands:

    a. The environment for the creatures on the Galapagos Islands remained unchanged for an extreme long time. They apparently have not encountered any danger the whole time. The physical part of their lives is therefore very simple. Accordingly, they could get by with very little technical advancements. The amount of spiritual advancements they need to match their technical advancement is thus also very little.

    b. Whatever imbalance they might have in their knowledge about life is therefore very small, which means the amount of distrust and fear they have is very small. Therefore, the imbalance doesn't get in the way of their being able to carry out their lives successfully.

c. Consequently, their creature nature did not develop much, if any, urge to fight or flight. Or if they had any initially it faded to where it is essentially nonexistent.

d. Members of their species residing elsewhere on Earth would have the fight or flight urge. This indicates that the environment these creatures are in for an extended period could shape their creature nature.

2. **Domesticating animals:**

a. Some animals could be more easily domesticated than others. This suggests that the creature nature of some animals is more easily changed than that of others.

b. Domestication could be perceived as a process of changing their creature nature by keeping them in an environment for a very long time where their behavior is constantly being influenced by humans, and they are being fed and cared for by humans.

3. **Domesticated creatures released into the wild:**

a. When domesticated creatures are released into the wild, their creature nature would revert back many generations later to how it was before they were domesticated.

b. The physical appearance of some creatures could also change. Pigs are an example.

An indication that the environment could shape human nature as well is that people from different parts of the world could have different human natures. However, I see a difference between creature nature and human nature. For one thing creatures would not do the horrible things to members of their own species that we humans would do to members of our own species.

Creature natures are neither good nor bad. It is what it is in order for their species to be viable in their environment. But the current state of human nature in my opinion is both good and bad instead of being neither good nor bad. I conclude it is because we humans have big egos. Our high level of intelligence, imagination, creativity magnifies both the good and bad in our current state of human nature. Thus, we could come up with wonderful ways to care for members of our own species and we could also come up with horrible things we could do to members of our own species.

We try to work around the bad part of our current state of human nature by establishing laws and rules, etc. These work to some degree, but our overall human behavior is still bad. In a way, I think laws and rules keep the bad part of our current state of human nature going because we humans tend to like to take on challenges and to win. The "challenge" here is how to break the laws and rules, and the "win" here is getting away with it.

But, our current state of human nature would have us want to establish laws and rules. It could be an "ego" thing in that the individuals establishing the laws and rules could feel powerful by doing so. In response, our current state of human nature would also perceive laws and rules to be like cages that are imposed on us from outside of us. Our current state of human nature would thus naturally want us to find ways to break out of the cages, and if we succeed, we would consider that to be a win.

This strange battle goes on within us because of our current state of human nature and our large ego. This indicates we need to come up with a different and better way to work around the bad part of our current state of human nature.

A better way would be to try changing our current state of human nature. We need to let the forces for change come from within ourselves instead of from outside of our selves. One way is to help us humans become more knowledgeable about life. As mentioned earlier we humans are now quite knowledgeable about the technical part of life but not very knowledgeable about the spiritual part of life. Therefore, there is definitely something we could do, and it is to help us humans become more knowledgeable about life.

Our current spiritual models such as religions, traditions, customs, etc. are able to do only so much, and it has not been enough. The problem is these models are not convincing enough, especially in today's high-tech way of life and with the population being more educated. They need to be more complete, and they need to be updated to increase their appeal in today's world. Everything else in our life gets updated periodically to maintain their appeal and therefore their effectiveness, so why not our spiritual models. Therefore,

1. The spiritual models need to be inclusive of technical considerations and also be updated to be more relevant to today's world.

2. The technical models need to be inclusive of spiritual considerations, especially those that are relevant to today's world.

Right now, none of the existing models has been reshaped in this manner.

If both kinds of models were reshaped in this manner and thus be more relevant and more complete in terms of modeling the two worlds we live in, they would be more convincing. Thus, the models would be more effective in helping us human become more knowledgeable about life because we human could relate to the models better. Again, this would be especially true in today's high-tech way of life and with the population being more educated.

By being more knowledgeable about life, the forces that could address the bad part of our current state of human nature would come from within us instead of from outside of us. These forces would thus be a part of us instead of being imposed upon us. Our current state of human nature would thus have a better chance of being changed for the better.

3.2. A Leader's Poor Behavior Could Be Sustained by Our Current State of Human Nature

We humans tend to make a bigger mess than usual when it comes to international relations. Leaders of nations shape international relations. How well or poorly they behave will affect the relations they establish. A few extremely poorly behaved leaders always seem to exist such that international relations as a whole tend to be messier than usual.

A nation that is among the most powerful ones needs to have good working relations with other nations. No nation is truly self-sufficient in today's world in which a wide range of capabilities and resources are needed to maintain its strength, and no one nation has everything it needs. A messy state of international relations is not going to make this interdependency among nations work well. In my opinion, our current state of human nature is not up to the task of assuring good international relations be consistently achieved or maintained, because it is currently such that it could sustain a leader who is behaving poorly.

Examples of our nation's lack of self-sufficiency include the following:

1. We rely on other countries, particularly China, for the rare-earth elements needed to build high-technology devices, modern aircrafts, hybrid and electric vehicles, etc.

2. The oil we produce by the fracking process is too low grade for our use. Therefore, we export it, and we import higher quality oil from other countries.

3. Much of our economy rely on the low-cost labor available in other countries.

4. We also rely on other countries for much of our manufacturing needs. Consequently, our expertise and manufacturing capability in many areas have now shrunked.

Our current state of human nature enables a leader who is behaving poorly to get away with behaving poorly in the following two ways:

1. **They do not have anyone above them to prevent them from behaving poorly, especially if they are dictators:**

    a. As evident in years 2017 through 2020 a leader does not have to be a dictator to get away with it if a number of people in powerful positions in the government and a large portion of the population support their leader's poor behavior because rightly or wrongly they consider it as a form of strong leadership, and they like strong leadership.

    b. Some members in government are fearful about what their leader would do to them if they oppose him or her because of their leader's would go after those who crosses him or her.

    c. A lot of people do not know the laws and rules of their governing system or how their governing system is supposed to work. Therefore, they are not aware their leader is breaking the laws and rules of the nation's governing system. This could potentially cause their democracy to change into being a dictatorship.

    d. Some people might feel disenfranchised by the nation's governing system and thus simply want change, any change, and thus they enjoy the disruptions their leader's poor behavior is causing.

    e. A lot of people tend to think short-term short-range much more than long-term long-range and therefore do not think about future consequences of leader's poor behavior.

2. **The attitude and values of the population keep changing from one generation to the next:**

The following is a brief recap of an earlier discussion:

a. In the beginning, the population is more into working together to be sure its civilization will thrive.

b. Each subsequent generation would know less and less about how their governing system is supposed to work and about its laws and rules.

c. Many generations later, the people now in charge are not concerned much about keeping their civilization thriving since it has been thriving for a long time, and it feels as if it will always be thriving, even as evidence exists to indicate the civilization is gradually declining.

d. The current generation is more into self-interests such that their civilization no longer receives the level of support it used to receive. Some members seem willing to ignore the laws and rules of their governing system for their self-interests. Perhaps it is because subconsciously they feel their governing system might be ending soon and they want to grab what they could before it ends.

The current state of human nature has a role in every one of the listed situations. The conventional approach is to impose laws and rules in an attempt to control the current state of human nature. But as explained earlier this only keeps the current state of human nature going. We need to come up with ways of improving the current state of human nature instead. One way is to for mankind to become more knowledgeable about life such that the will to improve would come from within us. As discussed earlier this would work better than imposing laws and rules upon us from outside of us.

3.3. Long-Term Long-Range Thinking vs. Short-Term Short-Range Thinking

Often it seems people do not differentiate between long-term long-range thinking and short-term thinking as if there only one way or the other to think. This can cause misunderstandings because both ways can right and yet be opposing.

1. **Our current state of human nature would have us think short-term short-range much more often than long-term long-range:**

    One of the problems we humans have is we often don't seem to take into consideration both long-term long-range and short-term short-range effects together and weigh them both before making our decisions.

2. **Long-term long-range actions and short-term short-range actions could both be valid and yet be opposing:**

    One of the likely reasons that we humans tend to not do this is because both long-term long-range and short-term short-range effects are valid. But both could also be opposing. We humans tend to think short-term short-range. Thus, we would tend to simply go with short-term short-range effects since they are valid and not think about long-term long-range effects.

3. **Real-life examples:**

    a. A real-life example is going on in years 2017 through 2020 in our nation. Members of one political party tend to think short-term short-range, and members of the other political part tend to think long-term long-range. The economy is doing very well, and that is good enough for the members of the political party that tends to think short-term short-range to excuse powerful members of government who are in their party who violated laws and rules of our governing system. Meanwhile the members of the party that tends to think long-term are having battles with members of the other party. In their opinion the current economy is created to be good by borrowing from the future economy and thus almost assuring a bad future economy.

The position one party is just as valid as the position of the other party. It all depends on whether you want a good current economy or a good future economy. However, the party that tends to think short-term short-range has two advantages. The population tends to think short-term short-range. The current good economy is right in front of us and is thus very convincing whereas the future economy whether good or bad is not right in front of us and is therefore not nearly as convincing.

Consequently, the party that tends to think short-term short-range is going to have an easier time winning the population's support than would the other party. This could explain why the economy is said by some people to have a cycle of its own. However, in my opinion, the cycle does not happen all on its own. I think it is also because we humans tend to think short-term short-range more than long-term long-range, and some politicians tend to cater to the population tendency to think short-term short-range.

This is another case of how the current state of human nature could contribute to making a mess on Earth.

b. Another example of mankind's tendency to think short-term short-range more than long-term long-range is in March of year 2020 when the world is dealing with the Covid-19 virus pandemic. Orders were given for everyone to shelter in place and to maintain a minimum of six feet separation between individuals. Going out is allowed only for essentials such as getting food.

At the time it seemed that young people, college age and younger, were thought to be much less likely to get severe symptoms or to die if infected, although this later turned out to be not quite true. Some young people thinking they are "safe" would hold parties, have large gatherings, and ignore the orders in general. They are not considering that they would be carriers of the virus if infected and could then spread it to those who are more vulnerable and could die. This includes those with preexisting conditions and people sixty years and older. They were obviously thinking short-tern short-range more than long-term long-range.

Incidentally, this second example is also an example of how when rules and laws are imposed upon us to control our behavior, we humans tend to look for reasons and ways to break the rules and laws if we think we could get away with it. It doesn't seem to matter that the rules and laws are meant to be for our benefit.

3.4. Steering the Economy Is Like Steering a Huge Ocean Liner at Sea

Unfortunately, some members of the population tend to not think long-term long-range enough to understand that steering the economy is like steering a huge ocean liner at sea. The economy changes course very slowly after action is taken to change its course. Therefore, a short attention span and a short-term short-range way of thinking would be at odds with anything that has an inherently long response time.

Consequently, those in the population who think short-term short-range are likely to make inappropriate decisions, especially when deciding their votes, because their way of thinking is at odds with reality. Such a mismatch is likely to create a mess without those who are creating it to realize what is happening.

Conversely, if a person in office takes action to assure a better future but causes a less comfortable present, he or she might not get reelected. The person later in office then gets the credit for the better future and is thus likely to get reelected. Thus, the person trying to do the right thing might not get reelected while the person who did nothing is likely to get reelected. This could cause a mess, and the voters creating it are not likely to realize their actions caused it.

Both cases tend to downgrade the quality of the kind of people we put in office, and this too would contribute to making a mess.

3.5. Take a Partnership Attitude and Not an Opponent Attitude

Whenever we come face to face with something alive or dangerous that we don't know much about we are likely to feel distrust and fear ranging from a negligible amount to a large amount depending on the situation and what that something is. Distrust and fear will trigger an urge to fight or flight ranging from a negligible amount to a large amount depending on the amount of distrust and fear.

Our life consists of a physical part that takes place in our physical world and a spiritual part that takes place in the spirit world. The physical part is about the physical things that we live with and work with. The spiritual part is about our direct and indirect interactions with living and nonliving things; i.e., it is the part in which we form our thoughts, decisions, intentions, solutions, goals, etc. Therefore, the spiritual part of life is more about how we go about carrying out life. Currently, we know a lot about our physical world and not much about the spirit world. Consequently, we know a lot about the things we live with and work with but not a whole lot about our direct and indirect interactions with living and nonliving things.

This means we know only about half of what we ought to know about life in order to be able to carry out life with plenty of wisdom. Because a large part of life exists that we don't know much about we tend to have distrust and fear as we go through life, even if the distrust and fear is only a tiny bit. This could explain why we tend to have an urge to fight or flight almost constantly, even if it is only a tiny bit.

In my opinion, this is why some individuals tend to have the following traits:

1. Judgmental of others and concerned about being judged.

2. Concerned about status.

3. Greedy.

4. Controlling.

5. Low in empathy and compassion.

6. Tend to see others as opponents.

7. Being very tense speaking before a crowd.

It is mainly because we can see and feel that we are all separate entities. But if everyone were to realize we are a part of one another in the spirit world and that part of life goes on forever, then we would understand we are only separate entities in our physical world for a purpose, and that purpose does not include harming one another. We would then be more into working around our separateness and toward fulfilling mankind's primary purpose for being here on Earth.

The traits listed above would then greatly decrease and could even go to zero if every one of us is as knowledgeable about the spirit world as we are about out physical world. We would then have the following traits:

1. Everything we do with others we would do it with a partnership attitude. We would tend to see others as partners and not as opponents.

2. We would value diversity and be open to how diversity could help make things more complete and therefore better. In life, diversity in general has helped make things be more complete and therefore better.

3. We would do something for its contribution to something worthwhile rather than for status. Status might come along but it would be secondary.

4. Perceive being old as being the time when we have the potential of making very meaningful contributions to mankind because of the knowledge we have accumulated. Making our contributions to be our project would tend to keep us young in our outlook on life. This because we would have a real meaning for being alive and we would thus want to stay alive.

3.6. Our Actions Tend to Be Locked In by Our Institutions and They Do Not Match the Need of the Spirit World

According to the spiritual model presented in Reference 1 mankind was brought into being on Earth to help restore balance the spirit world's state of knowledge. I think everyone has the feeling we humans have a purpose for being here on Earth, but we have not been able to figure out what it is.

Our consciousness is designed to be more tuned into our physical world than to the spirit world. Thus, we are not likely to know much about the spirit world and even less that we are carrying out our life partly in the spirit world and partly in our physical world. Consequently, we are almost certain to conclude that our purpose is to accomplish something in our physical world for the sake of our life in our physical world.

While it is true that our purpose is to accomplish something in our physical world, but it is supposed to be for the sake of the part of our life that is being carried out in the spirit world instead of for the sake of the part that is being carried out in our physical world. This misdirection of our efforts is understandable, but unfortunately it becomes very difficult to change because so many of the well-established institutions we humans developed are designed to be in support of the direction in which we are going on.

Such institutions include our schools, businesses, governing system, sports, entertainments, war industrial complex, etc. On a more personal level such institutions also include our traditions, customs, celebrations, games, contests, competitions, etc., and even partly in our religions. Meanwhile mankind's actions are mostly not matching the need of the spirit world.

What this means is we need to somehow modify our institutions such that our future generations would not be locked into acting is a manner that would not match the need of the spirit world.

3.7. The Urge to Fight or Flight

Mankind has a significant imbalance in its knowledge about life. It knows a lot about the part that is in our physical world but not much about the part that is in the spirit world. But it is the part in the spirit world that enables an individual to interact well with another individual and enables both individuals to feel comfortable with each other. It is the part where we choose our attitude, make our decisions, form our thoughts, and find our kindness, empathy and compassion.

By being lacking in our knowledge about the part in the spirit world we are likely to always feel some amount of distrust and fear and therefore always some amount of fight or flight whenever we meet an unfamiliar individual. By analogy, it is like trying to use an unfamiliar complex piece of equipment without knowing much about it. We are bound to feel some distrust and fear about using it until we know enough about it by having used it numerous times.

This means if every individual on Earth were to get well in touch with his or her part of life in the spirit world, we would all be able to interact among one another wonderfully. Mankind's overall behavior would be greatly improved. Mankind would no longer be making a mess on Earth, and it would be able to clean up the mess it had made. Thus, we would have collectively improved the state of mankind's human nature.

However, instead of putting effort into getting well in touch with the part of life in the spirit world, mankind got trapped by the obstacles that are a part of our physical world, as explained in Chapter Six. Because of this, our attention remained focused on our physical world for thousands of years such that we eventually forgot we were supposed to get well in touch with the part of life in the spirit world. Consequently, we continue to have the urge to fight or flight.

Therefore, to satisfy this urge we turned our attention toward doing things such as:

1. Make competition a part of almost everything we do, and going for winning.

2. Perceive something or someone as an opponent for a lot of things we do.

3. Participate in, or be an observer of games, contests, and awards programs that have win-lose outcomes.

4. Define conditions that could be indications of status and power, and be concerned about our own status in its many forms.

5. Design video games with violent actions with win-lose outcomes with lots of shootings and explosions.

6. Make movies and TV shows that are commonly to do with fights with criminals, the use of guns, explosions, wars in its many forms.

7. Compose stories about outer space adventures that are mostly to do with wars and battles with living things on other planets.

8. Etc.

Such actions are not necessarily totally good or totally bad. The fact that our physical world was designed such that living things are all physically separate entities indicates actions such as those listed were meant for us humans to do in our physical world. And, the kinds of new pieces of knowledge such experiences generate would include what the spirit world needs to help restore balance in its state of knowledge, provided they meet a certain requirement.

The actions listed could be bad in nature if we pursue them with bad intentions in mind, and they could be good in nature if we pursue them with good intentions in mind. For example:

1. Every one of the actions listed could be handled such that they would result in technically win-lose but spiritually win-win. In this case they would provide us with positive learning experiences for everyone involved for the benefit of everyone involved by leaving everyone's ego intact.

2. On the other hand, if they were handled such that they would result in technically win-lose and spiritually also win-lose, then the results would include bruised ego, anger, resentment, ill feelings, etc.

The first case is more likely to happen if we are equally knowledgeable about both the spiritual part of life and the physical part of life. The second case is more likely to happen if we do not know much about the spiritual part of life but we know a lot about the physical part of life.

It would be totally bad if our urge to fight or flight would have us humans do such as the following:

1. Make wars and battles of all kinds.

2. Pick a fight with mankind by doing criminal acts.

3. Go for self benefit at the expense of others or at the expense of the nation.

4. One group of people mistreating and/or hating various other groups of people.

5. Bullying people.

6. Hunt animals and other creatures for the sake of trophy hunting.

7. Persistently go for win-lose in just about everything we do.

8. Etc.

The main point in discussing the actions listed is that we need to be balanced in our actions based on a balance in our knowledge about life.

For example, participations in games need not be always win-lose. Little league sports should be win-win spiritually even if they could be perceived as win-lose technically. The spiritual win-win should count way more than the technical or physical win-lose.

Such a separation of the spiritual part from the technical part could be made, and letting the spiritual part count way more than the technical part could be applied to just about everything listed. This is an example of what I mean by being balanced in our knowledge about life.

By doing this, then, for example, the actions such as those listed could be altered in such a way that they would not be "fight or flight" in nature but instead could be learning experiences that could help us learn how to improve the quality of our interactions personally, nationally, and internationally. This would help make mankind's life on Earth to become closer to how life is in the spirit world.

Unfortunately, the way we are handling such actions causes so much complications that we tend to not put much effort into getting ourselves out of the trap we got caught in or into trying to get well in touch with the spiritual part of life. And by not being well in touch with the spiritual part of life, we are unable to eliminate the imbalance we have in our knowledge about life. Therefore, we are unable to form the wisdom necessary to figure out how to get our selves out of the trap. It becomes a vicious circle that has been going on for thousands of years.

What this means is that in order to get us out of this vicious circle we need to do something unconventional. Conventional thinking has proven to be ineffective.

3.8. Some People in Government Behave Like an Adult Gang

Terry Lewis does a lot of work with gangs, Reference 6. He works with young people, but in my mind what he learns about young people and their gangs could also apply to adults and their gangs. I think the current state of human nature is pretty much the same for the young people he works with and for people typically in government. A difference is adult behavior is generally less out in the open and is thus less likely to be recognized as gang behavior.

Also, adult gangs exist in more formal settings such as work places and government organizations. Because of these settings, we are more likely to accept this behavior as a way of doing business even if an adult gang is operating in a manner that could be considered criminal. Examples would include some dictatorships in which some of the members in the government are behaving in criminal ways, but we continue to maintain a relation with them because it serves us economically. A situation could also exist such that if the dictatorship were overthrown a worse governing group would take its place. In such a situation we are likely to continue to support the dictatorship now in power even if the governing group behaves like an adult gang. However, what is awful is when we are helping to prop up such a group with our economic dealings with it.

One of the features of young people gangs described by Terry Lewis is that getting in is easy and often encouraged as a means for survival, but trying to get out would usually result in a fatal consequence. An analogous situation exists within a certain governing system. At the start of his first term in office, the leader was reported to have said to members in government, who are of the same political party he is in, that they need to be loyal to him or he would find people among his numerous acquaintances to run for office and defeat them in the next election. This is disturbing in the same way that a young person trying to leave a gang might result in a fatal consequence.

For this and other reasons the general impression is that this leader is treating members in government, who are of the same political party he is in, as if they are members of an adult gang. And the amazing thing is those members are willing to follow as if they are members of an adult gang. This situation points out how fragile the governing system is. Its nature could be significantly changed easier than expected. All it takes is a leader who is willing to break the laws and rules of the governing system and a few individuals in very powerful positions in the governing system to support him to result in a change. This certainly should be a "wake up call" for the nation involved.

A democratic governing system is fragile. A major reason it is fragile is because we humans have two incompatible desires as explained in Chapter Two, Section 2.6. The result is we are not giving complete enough support for our democratic governing system by not caring enough that certain members in government in powerful positions are breaking the laws and rules of our governing system. Such an initial break down of our governing system could only continue to grow and would be very difficult to repair.

3.9. Some People Need a Strong Leader Simply for the Sake of Having a Strong Leader

It has been said by experts in human behavior that some people simply admire a strong leader, and it doesn't matter whether the leader does good things or bad things or what it is he or she is leading about. The spiritual model presented in Reference 1 provides a possible explanation for this.

Before the physical part of our life in our physical world began, we had only the spiritual part of our life in the spirit world in the form of the spiritual entity that would serve as our spirit. Because of what the spirit world is made of and how things work in there, life in the spirit world would be as follows:

1. Everything is directly or indirectly a part of everything else.

2. Because every human spirit shares a part of itself with every other human spirit, every human is directly a part of every other human in the spirit world.

3. Communications are instantaneous within the spirit world and they would cover the entire spirit world.

4. Because every piece of knowledge that exists at any point in time would reside in the spirit world at that point in time, all communications would be as complete and clear as possible and be as accurate as possible at that point in time.

5. Because every human spirit is a part of every other human spirit, a leader is not needed to bring every human spirit together. All human spirits would simply do everything together. No one human spirit could possibly feel insecure, alone, any distrust, or any fear. Ego and the urge to fight or flight are only concepts in the spirit world and are not practices.

Once the physical part of our life has begun in our particular physical world, it would be as follows:

1. We are all separate physical entities in our particular physical world.

2. Communications are mostly not instantaneous, and they are usually more localized than global.

3. Because thoughts must be translated into a form that could be expressed with a spoken language, communications are hardly ever as complete, clear, and accurate as possible.

4. Because we are all separate physical entities in our particular physical world, a leader is needed to bring everyone together.

5. Because no one could ever be perfect either in the spirit world or in our particular physical world, no leader in our physical world could ever be perfect either.

6. Because of all the short comings that are a part of our particular physical world compared with the spirit world, some people are likely to really need a leader in order to feel secure in our particular physical world.

This could explain why some dictators are able to maintain control of a nation, why some cult leaders are able to have a large following, and why our current president is able to have a large base of followers. These leaders could be perceived as being strong leaders, and it matters less what it is that they are leading about.

Those of us who feel secure and do not feel lonely, distrust, or fear might not understand why those that feel strongly the opposite way would behave the way they do, such as be so willing to follow a cult leader. According to the spiritual model in Reference 1, the sharp contrast between how life is in the spirit world and how life is in our particular physical world could in a sense be quite unnerving for some people and could thus cause them to feel quite insecure, lonely, distrustful, and fearful.

All this is going on subconsciously from the moment they were born such that such individuals are not likely to realize the way they feel is any different from how anyone else feels. However, such individuals are more likely to need a strong leader for the sake of having a strong leader, and it might not matter what it is the leader is leading about.

3.10. A Leader Needs to Understand the Nation's Population

It always bothered me that most of the candidates running for office are individuals who have been very wealthy all their lives and have lived a wealthy lifestyle all their lives. How are they able to understand the concerns of the population that they want to lead and therefore how could they be effective leaders for the population? Some might say a leader could surround him or her self with good advisors who are not very wealthy. However, it seems that kind of arrangement doesn't happen very often. Wealthy individuals tend to have wealthy friends and acquaintances.

In my mind, ideally a leader needs to have had hands-on experiences with doing the things typical members of the population would do while carrying out their lives. This could include such things as mowing the lawn, painting the house, work on their budget, doing their taxes, making investments, maintaining their car and various household appliances, dealing with re-roofing contractors, interacting with school administrators, etc.

Someone who is wealthy enough to hire out all such chores is not likely to fully understand the concerns of the population. In addition, the more money a candidate is able to spend on his or her campaign the more likely he or she will be elected. This tends to enable a wealthy candidate more likely to be elected and thus the disconnection between the person in office and the population worsens.

Therefore, I think the voters need to take into consideration the history of the lifestyle of the candidates as one of the major factors in deciding their votes. Right now, this is not being considered or even mentioned during election campaigns.

3.11. Interactions with Intelligent Beings from Outer Space

Our urge to fight or flight as discussed earlier could explain why almost every story and movie about space travel and interaction with intelligent beings from outer space involves wars, battles, fights, and massive explosions. Such stories and movies are almost always popular and financially successful because they cater to our urge to fight or flight. The more spectacular the special effects the more they satisfy our urge. Battles and explosions as depicted in outer space can be far more spectacular than any that can occur on Earth. Entire planets could be shown being blown apart. Fantastic futuristic weaponry, space crafts, and various war making equipment could be shown to function in amazing ways. Not only could such visuals greatly satisfy our urge to fight or flight, they could also be fun to watch and inspiring for our imagination and creativity; provided we know they are not real.

Anything that caters to our distrust and fear (the basis for our urge to fight or flight) also tends to be financially successful. For example, commercials are always about why we should buy and use certain products. The message is that our doing so would decrease our distrust and fear about something regarding ourselves. Conversely, it is often to create distrust and fear in us by suggesting the way we are doing things is lagging far behind what the latest technology has to offer such that everyone else is way ahead of us and leaving us at a disadvantage to compete.

Stories and movies about space travel suggest intelligent living things on other planets are warlike and want to take over Earth. By contrast, the spiritual model presented in Reference 1 suggests that if intelligent beings on other planets are able to figure out how to space travel from one planet to another in a matter of seconds, minutes, or hours instead of decades and centuries, they are more likely to be friendly than warlike. If they were warlike, they would have spent a major part of their time, energy, and resources making and preparing for wars such they would have little time to figure out how to do space travel in the very advanced way that they are doing.

We humans on Earth are warlike and we have wasted a major part of our time, energy, and resources making and preparing for wars. That is why we have not figured out how to do space travel in any advanced way. We have only thought of doing it by brute force.

If we were to get visitors from another universe, as opposed to from another planet, they would be even more likely to be friendly than hostile. In addition to their not wasting time, energy, and resources making and preparing for wars, they have no reason to want to take over our particular physical universe. What we have in our universe would be useless for them because our universe would not be compatible with their universe. As discussed earlier, every universe is different and unique.

Taking all this into consideration, based on the spiritual model in Reference 1, any visitors from either outer space or from another universe would most likely be friendly, and they are here mainly out of curiosity and for gaining knowledge. This could explain why we see UFO's (unidentified flying objects) or UAP's (undetermined aerial phenomenon) once in a while for decades and why they have yet to indicate they are warlike or want to take over Earth or our entire universe.

According to the spiritual model in Reference 1, space travel from one planet to another in a matter of seconds, minutes, or hours is likely to be more possible by going through the spirit world than going through our physical world. This is discussed further in Chapter Twenty.

To know for sure if this is possible, we would need to understand what the spirit world is made of and how things work in it. We are also likely to become increasingly less hostile and warlike once we get seriously into pursuing spiritual advancements.

To prepare for our trip to the moon and back, we needed to understand the physical functions and capabilities of the human body so that we could appropriately design and develop the vehicle that would carry humans to and from the moon.

We need to do similarly to prepare for doing space travel or universe travel by going through the spirit world. We need to understand the spiritual functions and capabilities of the human spirit so that we could appropriately design and develop the spiritual entity that will in essence carry (carry in a spiritual sense) the human spirit to the spiritual form of another planet or of another universe.

Right now, a typical human spirit has distrust and fear and therefore the urge to fight or flight whenever it encounters any unfamiliar living thing. This has to be one of the factors that would prevent us from being able to find a way to do space travel or universe travel by going through the spirit world. After all, if living things on other planets or other universes are friendly and not warlike, our having the urge to fight or flight is not going to make any visit successful.

This means we need to first get rid of our urge to fight or flight, and this means we must gain an understanding of the spiritual part of life comparable with our understanding of the physical part of life. In other words, we need to evolve spiritually much further than we have, and it is up to us to behave in ways that would move our spiritual evolutionary progression along faster.

3.12. The Possible Formation of a Military Space Force

In year 2019, being considered was the idea of forming a military Space Force as an additional and stand-alone military branch in America. I think this needs serious examination as to its purpose, function, and responsibilities. I would not want outer space to become a war zone. Consider the following:

1. Outer space was declared long ago by all nations to be used only for peaceful purposes.

2. This enabled the space station to be jointly used by Russia and America for peaceful space explorations even though it was initially constructed by Russia.

3. This enticed commercial enterprises such as Space X to do peaceful space explorations besides government enterprises. I think the involvement of commercial enterprises in this manner encourages the use of outer space only for peaceful purposes, especially when such enterprises are talking about providing peaceful commercial travels into outer space as vacation trips.

4. But now in year 2019 the idea of forming a Space Force as a stand-alone military branch in America seems to be in preparation for future wars in outer space. For example, ideas such as knocking out "enemy satellites" are being talked about. In my mind, such ideas sure sound warlike even before a Space Force becomes a reality.

5. As the most powerful nation on Earth I think we ought to lead the way for keeping the use of outer space only for peaceful purposes and to convince other nations to also use outer space only for peaceful purposes. We should definitely not be leading the way to make outer space a war zone.

Chapter Four

The Spirit World and the Spiritual Model Presented in Reference 1

4.1. The Spiritual Model Presented in Reference 1

The search for my personal spirituality turned serious after my mom passed away three years following my dad's passing. Four strange events occurred shortly afterwards that convinced me that an afterlife exists after death. The messages from the four events did not come directly to me. Instead they came through people who are not family members. I think this was purposely done by the spirits of my deceased parents to make sure I would realize I did not imagine the messages because of my heightened emotional state at the time. I think they also knew I have been looking for strong evidence that an afterlife exists after death. So, they provided it. The four events are described in detail in Chapter Two of Reference 1.

Being an engineer in my profession and was thus used to figuring out how things work I was driven to figure out how our afterlife is connected with our life when we are alive in our physical world. My search resulted in my formulating the spiritual model presented in Reference 1.

According to the spiritual model the afterlife is the spiritual part of life that exists in the spirit world. For each of us, it exists in the spirit world before our physical life in our physical world begins. It continues to exist in the spirit world while we are carrying out our physical life in our physical world. And, it will still exist in the spirit world after our physical life in our physical world is over.

In other words, our afterlife is our spiritual life that exists forever in the spirit world. It has a beginning but it does not have an ending. It began when a large enough spiritual entity was formed in the spirit world that is capable of serving as our spirit, and thereafter our spirit would exist in the spirit world forever. Therefore, this means our spirit is our afterlife and it is also our spiritual life that exists forever in the spirit world.

Thus, while we are alive in our physical world our life would consist of a spiritual part and a physical part. Our spiritual part will continue to exist after our physical part is over, and then we could refer to it as our afterlife. We could also refer to it as our before-life and our concurrent life, because it exists before, during, and after our physical part of life. But it is more accurate and simpler to refer to it as our spiritual part of life while we are alive in our physical world.

Writing the Reference 1 book took almost thirty years as many lengthy interruptions took place including my wife's terminal cancer, my own cancer, our children growing up and starting lives of their own, medical problems and major events happening with my parents and my wife's parents.

In formulating the spiritual model in Reference 1, rather than starting with existing spiritual models such as the various religions, traditions, customs, ways of life, etc., I decided to start from scratch and try something unconventional but still logical and rational. Being an engineer, I took into consideration engineering logic in this unconventional revisiting of spirituality, which is something that has never been done before. The resulting spiritual model is unlike any that exists up to now. It provides a more complete coverage of life than any other existing spiritual model could, and it does it in a manner that is more direct and explicit than could be done by any other existing spiritual model.

Reference 1 in paperback is the second edition of the book. It corrected the typos in the first edition, and it has a "Preview/Index" that is not in the first edition. Feedback from readers of the first edition indicates it was difficult to read, because it is very dense with unconventional concepts. Therefore, a "Preview/Index" is added to the second edition to explain many of the unconventional concepts as well as to serve as the index for the book. A reader could look over the Preview/Index to get a preview of the book to help make reading the text easier.

Chapter Three of Reference 1 explains the spiritual model and its formulation process in great detail. It is the longest and the most difficult chapter to read. I realized this from the start. Therefore, I included here and there in that chapter some humorous conversations between my left brain and my right brain to lighten things up.

One of the reasons I decided to write this book, the one you are now reading, is to present the spiritual model in an easier to read format and to leave out most of the formulation process. Therefore, it might be possible to read this chapter of this book instead of Chapter Three of Reference 1 and still gain a reasonable understanding of the spiritual model.

The following are some of the more important features of the spiritual model. For further details, please see Reference 1.

1. **The indisputable initial concept:**

    a. The indisputable initial concept that started the formulation of the spiritual model is as follows:

    - **Something, somewhere, somehow knows how to enable our universe and everything in it to exist.**

    b. Therefore, it is knowledge that enables things to exist. This makes logical and rational sense.

    c. We might then ask: where did the knowledge come from?

2. **Knowledge is generated by experiences:**

a. We learn from experiences. Therefore, experiences must generate knowledge. This also makes logical and rational sense.

b. Knowledge must come in pieces since knowledge generated from several different experiences could be pieced together to form a reasonable understanding of something in addition to the things with which those experiences were involved. This also indicates that an experience would generate multiple pieces of knowledge.

For example, as a kid I like to take apart old windup tick-tock clocks that no longer work. By seeing how a clock is put together with tiny bolts and nuts and a bunch of different sized gears and springs, I learn how the minute hand and the hour hand could move at different rates, how the pendulum works to enable the clock to keep good time, and I learned how to use screw drivers and wrenches.

Then when I owned my first car, I was able to follow the mechanics manual to figure out how the engine and transmission work, what to do keep the car in good running condition, and how to make repairs when needed. Thus, all my cars are able to lasts over 250,000 miles. Right now, I own three cars, all of them over 40 years, all in good running condition, and all in good physical condition. One of them is in near show room condition. I love driving them because they have become one-of-a-kind cars.

c. Knowledge from one experience could, by analogy, provide guidance on how to configure a new experience for the purpose of generating certain other kinds of knowledge. Therefore, this is additional evidence that an experience must generate multiple pieces of knowledge. Some pieces of knowledge would be relevant to many other things besides being relevant to the thing with which the experience was involved. Once generated, a piece of knowledge lasts forever.

d. Among the multiple pieces of knowledge generated by an experience, some would be new and some would have been generated before. But only one copy of any piece of knowledge is needed by the spirit world. Therefore, duplicates would merge together and only one copy would remain. By analogy, one copy of a piece of knowledge could be used an unlimited number of times to form spiritual entities just as one particular word could used an unlimited number of times to form an unlimited number of sentences.

3. **A piece of knowledge is analogous to a word:**

   a. As suggested earlier, a piece of knowledge is analogous to a word. This analogy goes further as follows:

   b. A piece of knowledge is neither good nor bad. Perhaps except for a few exceptions, a word is neither good nor bad.

   c. A spiritual entity is made of multiple pieces of knowledge connected together. A sentence is made of multiple words connected together. Therefore, a spiritual entity is analogous to a sentence.

   d. A large spiritual entity is made up of multiple smaller spiritual entities. A paragraph is made up of multiple sentences. Therefore, a large spiritual entity is analogous to a paragraph.

   e. An even larger spiritual entity is made up of multiple large and smaller spiritual entities. An article, story, book, etc. is made up of paragraphs and sentences. Therefore, a very large spiritual entity is analogous to an article, story, book, etc.

   f. Only one copy of any piece of knowledge is kept by the spirit world. A dictionary lists only one copy of the meaning of any word. Therefore, in this particular sense the spirit world is sort of analogous to a dictionary.

Also, there is no limit to the number of new pieces of knowledge that could be generated. There is no limit to the number of new words that could be formed. Therefore, there is no limit to the growth of the spirit world just as there is no limit to the growth of a dictionary.

g. A spiritual entity is neither good nor bad. But, when translated into a form that could exist or be expressed in a universe, the translated form could be perceived as good in one universe and bad in another. A sentence could be neither good nor bad. But when expressed, it could be considered good in one culture and bad in another.

4. **Defining the spirit world:**

a. All existing pieces of knowledge reside in one place and only in one place. That place will be called the spirit world. This is appropriate because it is the place in which the spirits of all living and nonliving things reside. How it is the case is explained as the presentation continues.

b. Conversely, all existing pieces of knowledge together would form the spirit world.

c. Accordingly, any newly generated piece of knowledge would automatically be added to the spirit world. This is how the spirit world grows and learns.

d. The spirit world is constantly growing and learning because there is always something somewhere going through an experience. Therefore, new pieces of knowledge are constantly being generated.

e. The spirit world being a growing and learning thing is thus a living thing.

f. The spirit world could never be perfect or complete because it could never possess every single piece of knowledge that could be generated. New experiences are always possible; thus, new pieces of knowledge are always possible to be generated. There is no limit to the new pieces of knowledge that could be generated.

g. However, the spirit world would always be correct at any point in time, because it would possess all pieces of knowledge that exist at that point in time. This means the spirit world will always to correct but incomplete.

h. Because the spirit world will always be correct but incomplete, so would anything it creates be correct but incomplete.

i. However, this does not mean anything a living thing, such as a human individual, formulates would necessarily be correct, although it would be incomplete nevertheless. It has to be reasonably formulated in order to be correct, and of course it would also be incomplete.

5. **Formation of spiritual entities:**

   a. Every existing piece of knowledge automatically forms a connection of the first kind with every other existing piece of knowledge.

   b. Any newly generated piece of knowledge would immediately do the same.

   c. Two or more pieces of knowledge connected together would form a spiritual entity.

   d. Therefore, spiritual entities are spiritual things that are formed in the spirit world and exist in the spirit world. They are automatically formed because every existing piece of knowledge and every newly generated piece of knowledge would automatically form a connection of the first kind with every other existing piece of knowledge

e. This means everything that exists in the spirit world is in the form of a spiritual entity.

f. The largest possible spiritual entity that exists at any given point in time would be the spirit world.

g. We could appropriately say everything that exists in the spirit world is created by the spirit world since it is the combination of all the existing pieces of knowledge that exists at any point in time that makes up the spirit world at that point in time.

h. Some of the spiritual entities are capable of being translated into forms that could exist or be expressed in our physical world.

For example, some could be translated into a chair to exist in our physical world. Some could be translated into a thought to be expressed in our physical world.

It thus follows that we could also appropriately say our physical world and everything in it are created by the spirit world.

i. Because the spirit world is always correct but incomplete, every spiritual entity will always be correct but incomplete.

This also means our physical world and anything in it would at best be correct but incomplete.

6. **All spiritual entities are thought-like:**

a. Every spiritual entity consists of pieces of knowledge connected together. Some spiritual entities are thoughts. Thoughts are formed by certain pieces of knowledge connected together in certain ways.

When we want to express a certain thought, we would use our spiritual senses to locate the spiritual entity that is the thought we want to express. We would grab it and translate it into a form we could express in our physical world. The process goes very quickly such that we are able to carry on a conversation with others.

b. But every other spiritual entity is also certain pieces of knowledge connected together in certain ways. This indicates every spiritual entity is thought-like, even though they could be spiritual forms of things other than thoughts such as a chair, a table, a car, a human individual, etc.

c. Therefore, when a spiritual entity is translated into something that could exist or be expressed in our physical world, that specific something exists or could be expressed in our physical world because that specific spiritual entity is essentially thinking that specific something to exist or be able to be expressed in our physical world.

d. Interestingly, this supports what a scientist who is working to identify smaller and smaller subatomic particles once said that the smallest possible particle that exists in our universe seems to resemble a thought.

7. **The spirit world is essentially thinking our physical world to exist:**

a. According to the preceding discussion, which is based on the spiritual model presented in Reference 1, it is possible that our universe and everything in it exist because a bunch of spiritual entities in the spirit world is essentially thinking all of it to exist.

So, we better hope the spirit world doesn't get a headache and couldn't think.

(Hey, right brain, could this be like an example of how it is when a left brain and a right brain are conversing with each other like how they were doing in Chapter Three of Reference 1? Well, left brain, you should know. After all, only you would come up with such a ridiculous concept as two sides of a brain conversing with each other. No-no, right brain, you would be that ridiculous too.)

b. This would make even more sense when we consider an author composing a story and a reader reading that story.

The world of the story and everything in it exist because the author is thinking all of it to exist.

A person reading that story could then bring his or her own version of the world of the story and everything in it to existence by his or her thinking all of it to exist.

To go a step further, someone could make a movie or TV show based on the story by physically create a version of the world of the story and of everything in it to exist.

In all such situations someone is thinking a world and everything in it to exist. This is analogous to how the spirit world is thinking our physical world and everything in it to exist.

c. The world of the story is only imaginary in our physical world. Therefore, the imaginary experiences of the imaginary people in the story would not be real in our physical world. Thus, the experiences would not generate any pieces of knowledge in our physical world.

However, the people and the experiences in the story are real in the world of the story such that the experiences would generate pieces of knowledge in the world of the story.

Chapter 4

In other words, the world of the story would be within another universe, and it would be real in that universe. We in our physical universe could mentally enter into that world by reading the story but we cannot physically enter into it.

A major difference between the world of the story and our physical world is that the lifespan of the world of the story is quite short as determined by the author whereas the lifespan of our physical world is very long by comparison, and so far, it doesn't seem to have a set duration.

d. This discussion is fun to have. But it also tends to lend support to the validity of the spiritual model presented in Reference 1.

8. **Defining the spirit, soul, and spiritual expression:**

a. The spiritual entity that enables an individual to exist is serving as that individual's spirit.

The actual spirit is embedded within that spiritual entity; i.e., it is a portion of that spiritual entity.

It is the spiritual entity that reincarnates. The spirit does not reincarnate. This is why the spiritual entity has to be larger than the spirit. It is because the spiritual entity would serve as the spirit for a different individual or a different living thing each time it reincarnates. Therefore, it is employing a different portion of itself each time it is serving as a spirit for some living thing.

After the life of a living thing in our physical world is over, its spirit does not vanish. The spirit and the soul would stay together and form the spiritual expression of the deceased living thing. The spiritual expression would then reside forever in the spirit world as the afterlife of the deceased living thing.

b. Every piece of knowledge making up the spirit has a connection of the second kind going from that piece of knowledge to the physical body of the individual.

The combination of all the connections of the second kind forms the soul of the individual.

The soul is what carries the spiritual signals from the spirit to the body to enable the body to exist in our physical world. The soul also carries the feedback signals from the body to the spirit. It is this back and forth transmission of these signals that enables the spirit to pilot the body.

It is also this transmission of signals that the spirit is able to restore the body during sleep. The restoration is to repair the wear and tear on the body caused by the strong and harsh signals from our physical world that impinges on the body while the individual is awake.

c. Everything discussed here applies to all living things on Earth in addition to us humans. An exception is restoration would not apply to those living things that do not sleep and do not need sleep. Usually, such living things have a very short lifespan, and they are designed to serve a certain purpose that could be fulfilled over a short lifespan. Insects are examples; i.e., I believe most of them do not sleep.

9. **Instincts and intuition, where do their messages come from?**

a. The portion of the spiritual entity that is not being employed as the spirit of a living thing would contain the knowledge that forms the spiritual signals that make up the instincts for the living thing.

This explains why living things have instincts and where the instincts signals are coming from, and why instinct messages seem to be specifically designed for the living thing that gets them.

b. Intuition on the other hand consists of the signals formed by the knowledge in rest of the spirit world that is not a part of the spiritual entity that is serving as the spirit.

This explains why living things have intuition and where the intuition signals are coming from, and why intuition messages tend to be more general in nature than are instinct messages. In other words, intuition messages are not as specifically designed for the living thing that gets them as are instinct messages.

c. Instinct and intuition messages seem to consistently contain a good amount of wisdom. This is because they originate from the spirit world, and the spirit world needs to maintain its ability to form good wisdom and always have good wisdom in order to stay viable. This is explained by the spiritual model in Reference 1.

In addition, we humans always tend to reach out to our perception of God for wisdom when we are in deep trouble. Any perception of God would be a portion of the spirit world so that the spirit world being a source of wisdom is nothing new.

However, the spiritual model in Reference 1 now could explain why instinct and intuition messages would contain a good amount of wisdom. It is because their messages originate from the spirit world.

d. This could also explain why managers who follow their intuition tend to be more successful than manages who do not. It is because intuition messages naturally contain a good amount of wisdom.

10. **The source of the spirit world's creative powers (from the spiritual model's standpoint):**

a. The sources of the spirit world's creative powers consist of the following:

- The power every existing or newly generated piece of knowledge has to form a connection of the first kind with every other existing piece of knowledge.

  This enables the spirit world to create at any point in time all the spiritual entities that are possible to create at that point in time.

- The power every piece of knowledge making up the spirit of a living thing to form a connection of the second kind going from it to the body of the living thing residing in our physical world.

  This enables the spirit world to bring into being all the living things that reside in our physical world.

- Knowledge is often said to be the most powerful thing that exists, and according to the spiritual model in Reference 1, this is true.

b. Another vitally important factor that plays a major role in how the spirit world uses its creative powers is "wisdom." The spirit world must know how to apply its creative powers wisely in order to maintain its own viability.

It must keep its state of knowledge reasonably balanced in order to form good wisdom. Good wisdom in turn enables the spirit world to know how to keep its state of knowledge reasonably balanced.

It becomes a vicious circle that could go unstable quickly once the state of knowledge is allowed to go unbalance beyond a certain level. Therefore, one of the major tasks the spirit world has is to keep tract of the state of balance of its knowledge and to take action to restore balance as necessary.

One possible way to restore balance is to design and bring into being a new universe that provides the kinds of experiences that could generate the kinds of new pieces of knowledge needed to restore balance.

11. **The spirit and soul will contain a record of the individual's life:**

    a. As the individual goes through an experience, the experience will generate pieces of knowledge. Some would be new and some would already exist.

    b. The pieces of knowledge that are new to the spirit and the spirit world would be added to the spirit and the spirit world.

    c. The pieces of knowledge that are new to the spirit but they already exist in the spirit world would be added only to the spirit. Thus, the spirit will in essence gain access to those pieces of knowledge already existing in the spirit world.

    d. The pieces of knowledge that already exist in both the spirit and the spirit world would not be added to either the spirit or the spirit world.

    e. The spirit and soul of an individual would contain a record of the life of the individual from his or her begin to the present by virtue of the pieces of knowledge and connections of the first kind added to the spirit and the connections of the second kind added to the soul.

    This is supported by the fact that we humans have auras. An aura could be seen and read by individuals who could see and read aura and who could then gain various information about what that person has been through so far in his or her lifetime.

    According to the spiritual model in Reference 1, an aura is produced by the spiritual signals that are transmitted back and forth through the soul. Therefore, when an aura is seen, it is seen in the spirit world with the spiritual senses of the individual who is able to see and read auras.

f. Everything that is described regarding us humans should theoretically apply also for any living or nonliving thing. Thus, it would be interesting to know if individuals who could see and read aura would be able to see auras surrounding living things other than humans and also if they could see auras surrounding nonliving things.

12. **Defining the spirit and soul of nonliving things:**

a. A common perception is that only living things have spirits and souls. The spiritual model in Reference 1 indicates this perception has to be untrue.

In order for a nonliving thing to exist in our physical world it has to have a spirit and a soul just like how a living thing has to have a spirit and a soul in order to exist in our physical world.

b. Nonliving things could go through experiences and generate pieces of knowledge just as living things could.

The major difference is a living thing has some control over the experiences it goes through whereas a nonliving thing has no control over the experiences that it goes through. Experiences simply happen to a nonliving thing.

c. Otherwise, everything that was said about the spirit and soul of a living thing would apply also to the spirit and soul of a nonliving thing.

d. The same goes for the spirit and soul of something that is in between being a living thing and a nonliving thing. Examples of such things are viruses, prions, a living thing in the process of dying and becoming a nonliving thing, etc.

13. **The opening statement in Reference 1 describing what is in Reference 1 is as follows:**

    Spirituality is revisited in a manner never done before. A measure of engineering logic is used instead of relying strictly on faith, which is the case for spiritual models that exist up to now.

    The new spiritual model describes the inner workings of the spirit world, defines our spirit and our soul, and describes the role of our spirit and soul in enabling our physical body to exist in our physical world. Thus, it shows how we are carrying out our lives simultaneously in the spirit world and our physical world.

    For example, our spirit resides in the spirit world while our body resides in our physical world. We form our thoughts and decisions in the spirit world, and we put them into action in our physical world.

    The model can plausibly spiritually explain common everyday observations and experiences. Examples are:

    a. Why we need sleep.

    b. Why dreams tend to be surrealistic.

    c. Why evolution is spiritual first and physical second, how missing links can happen.

    d. How identical twins can communicate telepathically.

    e. How pets could understand their owners' thoughts telepathically.

f. How some individuals can see and read auras.

g. Why our universe was brought into being.

h. How quantum superposition and entanglement can happen.

Many other examples are also presented.

Such extensive correlation with reality means the model is valid to a large degree. No other spiritual model that exists up to now could do similarly, and do it in the technical sense this model could.

Thus, the model offers a new and very different perspective regarding spirituality. Ultimately, the model is capable of indicating why we are here and the purposes we are to fulfill.

4.2. The Workings in the Spirit World as Modeled by the Spiritual Model

The concept of an existing spirit world has been around for a long time. But exactly what it is has always been vague. It is generally envisioned as the place where God, Jesus, angels, etc. reside. The concept of heaven as possibly being a place in the spirit world sort of fits, but then is hell also a place in the spirit world? It doesn't seem likely. So, not only is the concept of an existing spirit world vague, it could also be confusing, particularly if God is supposed to be kind, loving, forgiving, and the creator of everything. Then why would a place like hell even exist at all?

The spiritual model in Reference 1 suggests "hell" is only a concept in the spirit world and is not an actual place. And, possibly "heaven" is also only a concept. The spirit world itself could be perceived by some individuals as being heaven, and that might be reasonable, depending on what would qualify as being heaven in some people's mind.

According to the spiritual model in Reference 1, the spirit world is not simply a place. It is a living, growing, learning, conscious, and intelligent spiritual entity that encompasses all religions, all concepts of God, and more. Each religion is a certain portion of the spirit world, and each perception of God is a certain portion of the spirit world. Each such portion is different and unique. This means the spirit world is larger than all religions and all perceptions of God combined.

This also means each religion is as valid as any other religion, and each perception of God is as valid as any other perception of God. Therefore, it is senseless to have battles and wars over religious difference and over differing opinions as to which religion is correct or which perception of God is correct.

Some of the more important features of the spirit world as described by the spiritual model presented in Reference 1 are already presented in the preceding section of this chapter. That section also talked about the spiritual model that is presented in Reference 1. Additional features of the spirit world and the spiritual model are as follows:

Note: Almost every one of the following features could indicate the spiritual model presented in Reference 1 is correct to at least some degree. As emphasized in this book and in Reference 1, any reasonably formulated model would be correct but incomplete. And as can be seen by the following features, the spiritual model in Reference 1 appears to be correct, but it is incomplete in that it would not be able to explain everything there is to explain. Some of the features could also partly reveal mankind's primary purpose for being here on Earth. A more complete description about our primary purpose for being here is given later as more about the spiritual part of life is presented.

1. **The source of the spirit world's creative powers (from the spirit world's standpoint):**

    a. This topic was covered in subsection number 10 of the preceding section, which is about the spiritual model in Reference 1 instead of about the spirit world.

For this reason, the subsection number 10 coverage of this topic is cited here since this section is about the spirit world instead of about the spiritual model.

There are a lot of commonalities between the coverage of the spiritual model and the coverage of the spirit world since the spiritual model is largely about the spirit world.

b. Therefore, for details please see subsection number 10 entitled **"The source of the spirit world's creative powers (from the spiritual model's standpoint)"** in the preceding section of this chapter.

2. **Knowledge is among the most powerful things that exist:**

    a. Words exist and could be expressed in our physical world because we find the appropriate spiritual entities in the spirit world with our spiritual sense and then we translate them into forms that could be expressed in our physical world.

    b. Words enable us humans to compose articles, stories, books, etc. and they also enable us to document, preserve, and communicate all of our technical advancements and spiritual advancements. Thus, words have been said to be among the most powerful things that exist.

    c. This is the case because they are translations of spiritual entities made of pieces of knowledge, and knowledge is among the most powerful thing that exists, and that is what the spirit world is made of.

3. **The spirit world creates everything:**

    a. The spirit world creates everything that is possible to create at any given point in time.

b. This is because everything that exists in our physical world could exist because a spiritual entity in the spirit world is enabling it to exist.

c. Everything that exists in the spirit world is in the form of a spiritual entity, and every possible spiritual entity that could be formed at any given point in time is formed by the spirit world at that point in time.

d. The largest possible spiritual entity that exists at any given point in time would be the spirit world at that point in time. Therefore, in a sense the spirit world formed itself.

4. **Things that spiritual entities enabled to exist:**

    a. Discussed earlier are the physical things that spiritual entities could enable to exist in our physical world and also the nonphysical things that they could enable to be expressed in our physical world.

    b. Spiritual entities enable feelings of all kinds to exist. Feelings are mental things that exist only in the spirit world, but we could feel them with our spiritual senses while we are in our physical world.

    Feelings are not things we need to find in the spirit world with our spiritual senses. They are a part of our spirit and they are triggered into action by our experiences and situations.

    It is like how our ability to feel something with our hands or other parts of our body when our sense of feel is triggered as our body comes in contact with something.

    c. Spiritual entities enable concepts of all kinds to exist. Concepts are also mental things and they also exist only in the spirit world. We are able to gain access to them with our spiritual senses and are able to understand them, describe them, and apply them in our physical world.

d. Spiritual entities enable designs of all kinds to exist. Designs are again mental things and they too exist only in the spirit world. We are able to gain access to them with our spiritual senses and are then able to copy them in a variety of ways in our physical world.

e. Spiritual entities enable mental states of all kinds to exist such as attitudes, intentions, our will, tenacity, determination, giving up, moods, etc.

These are also not things we need to find in the spirit world with our spiritual senses. They are a part of our spirit, and they are triggered into action by our experiences and situations.

f. Spiritual entities enable our spiritual senses to exist, and they also determine the acuity of each of our spiritual senses.

Our spiritual senses give us our imagination and creativity and our problem-solving ability, reasoning ability, critical thinking ability, etc. Such mental abilities involve our use of our spiritual senses to find things in the spirit world

g. The strength of such mental abilities might depend on the number of spiritual senses we have, their acuities, and our proficiency in using them. It is possible that each of us have a different number of spiritual senses, difference acuities in our spiritual senses, and different proficiencies in using our spiritual senses.

A person with a high IQ might have a larger number of spiritual senses than average.

h. A person with high "streets smarts" might have higher acuity in his or her spiritual senses than average and might also be more proficient in using his or her spiritual senses than average.

In my opinion, IQ and "streets smarts" are mental abilities that are separate from intelligence. However, high intelligence could boost both IQ and "streets smarts".

I know many people who have high IQ's who could thus get high grades in school and in college but they might not do so well in their careers. I also know many people who get average grades in school and in college who are highly successful in their careers because of their high "streets smarts".

5. **The larger the spiritual entity the more complex would be its translated form that could exist or be expressed in our physical world:**

a. I would speculate that moderately large spiritual entities could enable nonphysical nonliving things such as thoughts, concepts, stories, etc. to exist or be expressed in our physical universe.

b. Larger spiritual entities could enable physical nonliving things such as contemporary automobiles with all their computers and semi-self-driving features to exist in our physical world.

c. Very large complex spiritual entities could enable complex nonliving things such as giant airliners and giant cruise ships to exist in our physical world.

d. Extremely large complex spiritual entities could enable very complex and highly intelligent living things such as humans to exist in our physical world

e. Even larger and more complex spiritual entities could enable universes to exist.

f. The largest size and most complex spiritual entity that exists at any point in time would be the spirit world at that point in time.

6. **The number of spiritual entities would increase something like two times an exponential rate of increase with every new piece of knowledge generated and added to the spirit world:**

a. The number of spiritual entities would way more than double with every new piece of knowledge generated and added to the spirit world. Two processes are going on to increase the number of spiritual entities. Their total effect would result in a rate of increase that is two times an exponentially increase.

Process Number 1: Every already existing piece of knowledge would form a connection with the new piece of knowledge to form as many new two-piece spiritual entities as there are already existing pieces of knowledge. Therefore, a new crop of two-piece spiritual entities equal to the number of existing pieces of knowledge is formed with each new piece of knowledge generated and added to the spirit world. These new two-piece spiritual entities would join in with the rest to participate in Process Number 2.

Process Number 2: Every already existing spiritual entity would connect with the new piece of knowledge to spin-off a new spiritual entity composing of one piece of knowledge more than the initially existing spiritual entity. Meanwhile the initially existing spiritual entities would continue to exist. Therefore, the number of spiritual entities resulting from Process Number 2 would double. When the next new piece of knowledge is generated and added to the spirit world, twice as many spiritual entities would exist to be doubled as there were previously. Therefore, Process number 2 would result in an exponential rate of increase.

b. Meanwhile Process Number 1 has increased the number of

Chapter 4   133

two-piece spiritual entities by the number of already existing pieces of knowledge every time a new piece of knowledge is generated and added to the spirit world. A new crop of newly formed two-piece spiritual entities would also then be participating in the Process Number 2 exponential rate of increase every time a new piece of knowledge is generated and added to the spirit world.

c. Therefore, we have two processes are going on at the same time; each contributing separately to an exponential rate of increase. Thus, the total rate of increase is something like two times an exponential rate of increase.

7. **The spirit world could enable trillions upon trillions of things to exist in our physical world:**

The reason the spirit world could enable so many living and nonliving things to exist on Earth could be explained as follows.

a. To illustrate, let's begin with just one spiritual entity that could enable a living thing to exist on Earth.

b. Each new piece of knowledge generated and added to the spirit world would more than double the number of spiritual entities that could enable a living thing to exist on Earth.

c. First, all the spiritual entities that could enable living things to exist on Earth would double because each would form a new one that is one piece of knowledge larger than the original one and meanwhile the original one would still remain. Thus, the number able to enable living things to exist on Earth would double.

d. In addition, the added piece of knowledge would enable more spiritual entities to become large and complex enough to enable living things to exist on Earth. These then would then join all those that have been doubling with each new piece of knowledge generated and added to the spirit world, and they too would likewise be doubling thereafter. This is why the number able to enable living things to exist on Earth would more than double with each new piece of knowledge generated and added to the spirit world.

e. However, for simplicity let's focus only on the one spiritual entity initially under discussion that could enable a living thing to exist on Earth. Therefore, with one new piece of knowledge generated and added to the spirit world, that one spiritual entity would spin-off a second spiritual entity that could enable a living thing to exist on Earth. So, now we have two spiritual entities that could enable living things to exist on Earth, and thus we could now have two living things on Earth.

f. This doubling action would take place with every new piece of knowledge generated and added to the spirit world. Therefore, the number of spiritual entities that could enable living things to exist on Earth would increase exponentially with a continued generating and adding of new pieces of knowledge to the spirit world.

g. With just 30 new pieces of knowledge generated and added to the spirit world, the initial spiritual entity would increase to 1,073,741,824 spiritual entities that could enable living things to exist on Earth.

h. With just 50 new pieces of knowledge generated and added to the spirit world, the number grows to something like 1,152,921,000,000,000,000. These are just from the one spiritual entity under discussion and they do not include all the additional spiritual entities that grew large enough to enable living thing to exist on Earth. They too would have been doubling with each added new piece of knowledge so that even more additional spiritual entities would grow large enough to enable living things to exist on Earth.

i. Therefore, the actual number of spiritual entities that could enable living things to exist on Earth would be many times larger than the number given for after just 50 new pieces of knowledge are generated and added to the spirit world.

j. Consequently, we can see it doesn't take long before the spirit world is able to enable an extremely large number, and an extremely large variety, of living things to exist on Earth.

k. Keep in mind every living cell making up the body of a living thing, and every bacterium, is a living thing. Therefore, it does take an enormous number of spiritual entities that could enable living things to exist in on Earth to enable all the living things that exist on Earth to exist on Earth.

l. What is discussed about living things would apply in the same way to nonliving things.

8. **Continuously growing and changing long-living things need periodic adjustments:**

    a. Some things that are long-living, continuously growing, and changing are likely to need periodically attention and adjustments to stay balanced and viable.

    For example, a lot of plants are long-living, keep growing, and will change as new branches form and some old branches die or are pruned. Fruit trees and various landscaping plants are among those that need continuing periodic attention and adjustments.

    b. On the other hand, certain fishes are long-living and keep growing bigger, but they don't really change. Therefore, they would not need periodic adjustments.

c. The spirit world is forever living, growing, and changing. It is forever becoming more complex because every new piece of knowledge added to it is different from any already existing piece of knowledge. Thus, the spirit world at any point in time would always be more complex than it was at any earlier point in time.

d. The spirit world needs to keep track of the state of everything within itself in order to determine how to design the next universe it will need to help restore balance.

e. A lot of things exist within itself that need to be constantly monitored. Thus, a lot of experiences that are necessary for the spirit world to go through would be of a kind that could only be possible to exist in the spirit world. This means a lot of new pieces of knowledge the spirit world would be generated would be of a kind that could only be generated within the spirit world.

A constant stream of this kind of pieces of knowledge generated and added to the spirit world would eventually cause an imbalance to develop in the state of knowledge of the spirit world. Therefore, the spirit world would need to take periodic action to restore balance.

Such actions would often be designing and bringing into being a new universe to help restore balance. This would enable new pieces of knowledge to be generated that are different from those that could only be generated within the spirit world.

f. By analogy it would be like our needing to consistently spend time keeping everything in our home in clean, working, and organized. The experiences would generate a stream of new pieces of knowledge of the kind that could only be generated within our home. Therefore, to be balanced we need to have certain other kinds of experiences to generate certain other kinds of new pieces of knowledge that would counterbalance our home generated pieces of knowledge.

g. The spirit world would thus design and bring into being a new universe that has certain characteristics that will help the spirit world restore balance. Usually multiple causes of imbalance would exist such that the universe would be designed to counter the multiple causes.

h. Living things in the universe would not know what the multiple causes are. The universe would be designed such that if the most intelligent living things residing in it would fulfill their primary purpose, then all of the causes would be appropriately addressed.

i. Sometimes each of the multiple causes would be addressed separately in separate regions of the universe. This could be the case for our universe. Our universe is likely to have multiple planets that are hospitable to living things for this reason. On each such planet the most intelligent species of living things residing on it would be responsible for fulfilling the primary purpose of that planet. In the case of Earth, we humans would have this responsibility for Earth.

j. Every other such planet would have a different primary purpose. Therefore, the most intelligent species of living things in each such planet would likely to be different for each planet. This means we humans are most likely to be the most intelligent species of living things only on Earth. It will be interesting to find out what the most intelligent species of living things is like on other such planets some day when we have learned how to do space travel in a practical manner.

k. It makes sense to have just one species of living thing to be responsible for fulfilling the primary purpose for each such planet. Otherwise it could be confusing. Also, if there were more than one species responsible, they would have to stay in communication so closely that they might as well be of the same species. Having a single species be responsible would make more sense

l. A more detailed discussion about the specific primary purpose of any specific universe and what it has to accomplish to fulfill it is presented in Chapter Seven of this book.

m. In the case of our universe, in Chapter Seven, Section 7.6. is shown how the characteristics of our universe are in many ways counterbalancing the characteristics of the spirit world.

This indicates that our universe is likely designed such that among the new pieces of knowledge it is to generate, some would be of the kind that would counterbalance the pieces of knowledge the spirit world has generated within itself. The rest would include new pieces of knowledge that would counterbalance other causes of the imbalance.

n. In other cases, the new universe might or might not be designed to generate new pieces of knowledge that would include some that would counterbalance the pieces of knowledge the spirit world has generated within itself.

A new universe that is not designed to generate such new pieces of knowledge might be one that is nonphysical. This is because the characteristics of the new universe need not be opposite to the characteristics of the spirit world.

Therefore, since the spirit world is nonphysical. It would be OK to have the new universe to also be nonphysical. Living things might also not have to be separate entities as they are in our universe. Living things might also be very in touch with their spiritual senses, which is unlike how we humans are not very in touch with our spiritual senses. Etc.

Chapter 4 139

9. **Each new universe will be different from any that came before it:**

   a. The preceding discussion explains how it is the spirit world becomes increasingly complex as it grows. Therefore, it constantly changes and could never again be how it was at any time that has passed.

   b. This means whenever the spirit world needs to design a new universe to help it restore balance, the causes of imbalances would be different each time and thus could never be the same as they were at any such past occasions.

   c. This means each new universe could never be the same as any universe that has been designed and brought into being in the past.

10. **Multiple simultaneous universes and/or sequences of universes might be needed to restore balance:**

    a. Our universe is physical and three-dimensional. After our universe has done what it could to restore balance, the spirit world might then need to design and bring into being a universe with more than three dimensions, or it might need to be nonphysical, etc. to help make further adjustments.

    b. This could happen if the first universe did not fulfill its primary purpose or it somehow caused a new imbalance to form.

    c. It is also possible that the spirit world might to design and bring one universe into being after another after another because some of the imbalances were particularly difficult or tricky to rebalance and thus must be adjusted a step at a time.

    d. Another possibility is multiple universes might need to exist simultaneously, because the natures of the multiple imbalances are too diverse to be handled by a single universe. It is likely that our universe happens to be able to handle multiple imbalances well enough with multiple hospitable planets. However, there might still be other universes existing simultaneously with ours.

e. Living things in one multiple simultaneous universe might not be able to sense any of the other universes. The living things in one universe could have major senses that are able to sense only the attributes of its own universe and none of the attributes of any other universe.

We might ask, while we can understand why multiple simultaneous universes might be needed, but why couldn't a sequence of universes be avoided by making sure the living things would fulfill their primary purpose even if it is particularly difficult or tricky?

a. The answer is spirit world does not control the experiences the most intelligent living things residing in a universe would go through. This is because highly intelligent living things could form almost all of their experiences themselves.

b. If the spirit world were to be able to form all the experience for the most intelligent living things then it would already possess all the pieces of knowledge those living things would generate. Then the universe would not be needed.

c. Thus, it is possible for a universe to not completely fulfill its primary purpose and a subsequent universe would be needed. Or a universe might unintentionally cause a new imbalance to form. Then a subsequent universe would also be needed.

d. Earth appears likely to become a planet in our universe that is not going to fulfill its primary purpose, considering that we humans are not making much progress in improving our poor overall behavior. Therefore, even if the rest of the hospitable planets in our universe were to fulfill their primary purposes, because Earth failed, the spirit world might have to design and bring into being a new universe to do what Earth was supposed to do but failed.

e. The reason for my writing this book and Reference 1 is to bring about new and more effective ways for us humans to be inspired and motivated to improve our overall behavior and to eventually fulfill our primary purpose before time runs out on us.

11. **We humans do not create things, not even artists. But what we are able do could still be called our creativity:**

a. We often say "this person or that artist created this or that." But according to the spiritual model presented in Reference 1, and as stated earlier, everything that exists is created by the spirit world.

b. Therefore, when someone is said to have created something, that individual actually used what is called his or her creativity to find the spiritual form of that something in the spirit world and then translate it into a form that could exist or be expressed in our physical world. That individually did not create that something.

c. That person's creativity thus consists of being able to do the following:

- He or she is exceptionally in touch with his or her spiritual senses.

- He or she is very proficient in using his or her spiritual senses to find spiritual forms of things in the spirit world.

- He or she is also very proficient in translating that spiritual form into a form that could exist or be expressed in our physical world.

d. This does not mean a person's creativity should not be valued as much as before. It should be highly valued just as it has always been highly valued.

- Any individual who is exceptionally able to find things in the spirit world and is exceptionally able to translate such things into forms that could exist or be expressed in our physical world would be enriching our lives by broadening our experiences.

- When such an individual is working in a technical field, he or she is often exceptionally able to move forward our technical advancements.

e. For the same reason, an individual with exceptional creativity could similarly move forward our spiritual advancements as well.

f. Most importantly in both cases such individuals are helping us to be more aware of what is in the spirit world.

g. A reason why we very much appreciate fine arts is because they make us feel a bit like we are home again in the spirit world where we originated from in the first place.

h. This could also explain why a good story or a good movie could bring tears to our eyes. They could make us feel a bit like we are home again where we long to be.

12. **We could imagine or think about something only if that something is already created by the spirit world:**

There is a difference between imagining and thinking about something. When we are imagining something, we are exploring one or more possibilities. When we are thinking about something, we are examining something specific.

For example,

a. We can think of a specific shirt we saw in an ad (we found the spiritual form of that specific shirt in the spirit world), and we are going to the store to find and buy that specific shirt.

b. We saw in an ad that all shirts are on sale and we are imagining we might find something we like (we found the spiritual forms of a random bunch of shirts in the spirit world), and we are going to the store to see if we like any.

Another example:

a. We see an acquaintance we have not seen a long time and we want to say hello. Therefore, we would recall his name first (we find the specific spiritual form of his name first) before saying hello.

b. We want to say hello, but we cannot remember his name. Therefore, we are imagining a whole bunch of possible names (we are searching through the spiritual forms of a whole bunch of possible names to see if we could locate the correct one) to see if we could identify the correct one.

A third example:

a. We are ready to start our college education and we know what we want to major in. Therefore, we would find out all the courses and various other requirements involved.

b. We are ready for college but can't decide what to major. Therefore, we would explore a wide range of possible majors and then imagine which would fit us best.

These examples covered only one of countless ways we go about carrying out our life simultaneously in both worlds.

13. **We form our thoughts, decisions, attitudes, purposes, etc. in the spirit world with our spirit and we carry them out in our physical world with our body:**

When our life in our physical world is over, our body including our brain is deceased. But our consciousness, intelligence, memories, and the rest of our mental abilities all continue to exist, and they exist with our spirit. This indicates our spirit is where our learning process takes place, and it does not take place in our brain as commonly assumed.

We could see some changes taking place in our brain when we learn something, but it is not directly because we learned something. Our spirit did the learning and it thus makes some adjustments in our brain so that our brain could then interact with our spirit regarding what was learned. Therefore, the changes we see in our brain is an indirectly result of our spirit having learned something.

All of what is said makes a lot sense because our spirit is made of pieces of knowledge, whereas our body including our brain is not. It is the pieces of knowledge involved that are making all of our mental abilities possible, and it is our spirit that is made of pieces of knowledge. This means the following:

a. All our mental abilities reside with our spirit including our ability to form our thoughts, decisions, attitudes, purposes, etc.

b. We then carry out things such as these in our physical world with our body. How we do this is by our spirit "piloting" our body much like how a pilot of an airplane pilots the airplane or how a driver of a car pilots the car.

c. Our brain is essentially the computer center for our body:

- Our spirit sends spiritual signals through our soul to our brain.

- Our brain translates the spiritual signals into electrical signals it then sends to our body to get our body to move the way our spirit instructs it to move.

- Our body then sends feedback electrical signals to our brain.

- Our brain translates the feedback electrical signals into feedback spiritual signals it then sends to our spirit via our soul to let our spirit know how well our body is responding to its instructions and what condition our body is in.

- This back and forth sending and translating of signals is how our spirit pilots our body. The process goes extremely quickly.

d. Our spirit also has direct connections with the rest of our body besides our brain via our soul. This is how our spirit enables the rest of our body and its vital organs beside our brain to exist, be alive, and to function.

e. This direct connection between our spirit and the rest of our body is what enable our body to have its body intelligence. This includes a lot of the bodily functions of our various vital organs and it also enables our body to be able to do certain things very proficiently without going through our brain. This happens when our body has been doing this certain something for a long time. An example is how our body manages to keep a bicycle upright while we are riding it without our consciously thinking about keeping it upright. This happens after we have been riding a bicycle for a long time.

14. **The size of the spirit and the size of the brain of a living thing are not necessarily related to each other.**

a. For example, the human spirit is large because humans are highly intelligent and have a large amount of mental abilities.

The human brain is also large because the human body could do so many different things and could them with great precision.

b. On the other hand, some very smart birds have very small brains because their bodies could do only a very limited number of things.

Meanwhile their spirits are very likely to be much larger than we might expect based on the size of their brain. This is because very smart birds have a fairly high level of intelligence.

c. In other words, according to the spiritual model presented in Reference 1 and as reviewed here:

- The size of the spirit has to do with the level of intelligence and the number of mental abilities a living thing has.

- The size of the brain has to do with the number of things the body could do and the precision by which the body could do them.

- The size of the brain also depends on the size of the body it has to serve as the computer center for. Thus, the brain of a large whale will be quite large simply because the whale is large. The brain of a bird will be small not only because its body could do only a limit number of things, its body is also small.

- Therefore, what this means is the size of the spirit and the size of the brain are not directly associated with each other.

15. **Why do we humans need sleep, why do we dream, and why are dreams usually surrealistic?**

    a. Why we need sleep: This back and forth transmission of spiritual signal directly between our spirit and our body as described in the preceding Item 13 is what restores our body while we sleep. This is to restore our body from all the wear and tear that happens when it deals with the very loud and harsh signals from our physical world while we are awake during the day.

    - The need for regular restoration is why humans need sleep.

    - During sleep we go through several cycles of REM sleep.

      I think the reason for the cycles is because the spirit world realizes we often do not get the full amount of sleep we need. Therefore, in order to be sure that the most important parts and organs of our body get restored, the spirit world designed us to sleep in cycles.

    - The first cycle is most likely to restore our brain. The next cycle is most likely to restore our heart. Etc. This goes on in the order of priority set by the spirit world until our entire body is restored.

      If we do not get the full amount of sleep we need, then at least the most important parts and organs of our body got restored, and we are then able to get by for the next 16 or so hours while we are awake. However, we are not likely to be at our most efficient or would we be feeling our best.

    - This explanation of why we humans need sleep is most likely applicable to other living things as well such as animals and birds, etc.

It probably does not apply to most insects, because I don't think most of them sleep. This is probably why their lifespans are very short by comparison with animals and birds, etc.

b. Why we dream: During the restoration process, our spirit reviews the experiences we have gone through during the day in order to assess the wear and tear that our body has experienced. This enables the spirit to determine more precisely the restoration that is needed.

This review of our daytime experiences is what produces our dreams. Our dreams might not always be associated only with our latest experiences since some of the wear and tear might have something to do also with some experiences in the past.

c. Why our dreams tend to be surrealistic: The reason our dreams tend to be surrealistic is because our evolution process spans over multiple universes. Only the latest portion of our evolution process took place during our current lifetime on Earth.

When the spirit world is reviewing our daytime experiences to determine how best to do the restoration, the restoration is to bring the state of being of our body back to how it was before all the wear and tear happened during the latest daytimes. I say "daytimes" instead of just "daytime" because previous sleep durations were not long enough to enable previous restorations to be completed.

Incomplete restoration could be especially a problem for today's young folks who get addicted to their smart phones, students who have lots of homework, and seniors who simple do not get enough sleep due to various medical conditions.

The state of being that our body was at before the wear and tear took place was formed from all the evolutionary changes that the spiritual entity serving as our spirit has gone through over multiple past universes. These past universes are going to be different from each other and also different from our current universe.

This means our spirit is made up of a mixture of pieces of knowledge generated from all these past universes plus pieces generated in our current universe. The restoration process is going to involve this diverse mixture of pieces of knowledge. Consequently, the dreams the restoration process will produce are going to be a mixture of images from past universes and images from our current universe. This is why our dreams would usually have a surrealistic quality.

The notion that the spiritual entity serving as our spirit has been through multiple past universes tends to be supported by the fact that while our dreams are surrealistic, we don't seem to be surprised or mind that they are surrealistic while we are dreaming. We simple deal with the surrealistic nature of our dreams as if it is nature. This indicates that our spirit accepts the surrealistic nature because it knows it self is made up of a diverse mixture of pieces of knowledge generated from multiple universes.

This tends to confirm the correctness of the spiritual model presented in Reference 1. It also lends support to the notions that multiple past universes have come and gone and that the spirit world would periodically design and bring into being a new universe to help restore balance in its state of knowledge.

This also means if we were to develop a way to separate out the parts of our dream images that pertain only to our current universe, we would be left with images that pertain to past universes. This might provide us a way to gain some understanding of some of the past universes.

16. **We are able to find and grab thoughts quickly enough to carry on a conversation:**

a. The process described in earlier discussions applies to how we go about finding thoughts in the spirit world with our spiritual senses and then translating them into forms that could be expressed in our physical world.

b. The process works essentially instantaneously such that we are able to carry on a conversation.

c. At times we might have trouble finding the right thought to express in a difficult situation. When this happens, it tends to support the notion that we do find our thoughts somewhere somehow, and the somewhere is the spirit world and the somehow is with our spiritual senses.

d. Often, we would look for a thought or a series of thoughts starting from a central topic. It might be the topic under discussion with someone, or it might be a topic of an article we are reading, etc.

For example, suppose the conversation is about global warming. We would start off grabbing the spiritual entity that is the spiritual form of the term "global warming."

This spiritual entity would be a part of every thought that has to do with global warming, and thus all of them would be "right there next to our spirit" in the spirit world and would be very easy for us to grab and express.

e. Remember, the spirit world is nonphysical and does not need space in which to reside like how Earth needs space such as the space of our physical world in which to reside. Things are in essence "right next" to each other such that going from one spiritual entity to another can be done essentially instantly.

f. Therefore, when someone said something to us, that person sets in place the topic of discussion. We then grab the spiritual entity that is the spiritual form of that topic, and all the thoughts that have to do with that topic would be immediately accessible to our spirit.

We could then immediately pick and choose which thought to grab as our response. This is how we are able to carry on a conversation.

17. **We could do a variety of things because our spiritual senses function extremely fast. This even includes letting the spirit world do a bit of our thinking for us:**

    a. The preceding discussion indicates our spiritual senses function extremely fast. This explains why we are able to do a variety of things besides carry on a conversation.

    b. Some examples include:

    - Driving a car, pilot an airplane, etc.

    - Participate in all kinds of sports.

    - Doing what "first responders" do in an emergency.

    - Playing video games.

    c. This could explain why we could react extremely quickly and instinctively to something very dangerous before we even think about it.

    It is in this very limited sense that we humans still rely on the spirit world to do a bit of our thinking for us.

    More specifically, our spiritual senses function so much quicker than our major senses could that they and the spirit world could take over getting our body to react before our major senses and our spirit could do so.

18. **The text of a book must already exist in the spirit world before an author could compose it:**

a. We might ask, can something as extensive as the entire text of a book already exists in the spirit world? And, could this even be true for the entire set of an encyclopedia?

b. The answer is yes.

c. Remember that the rate of increase in the number of spiritual entities in the spirit world is something like two times an exponential rate of increase for every new piece of knowledge generated and added to the spirit world.

d. Therefore, if there are enough spiritual entities to enable everything making up our universe to exist, there is enough to enable the spiritual form of the text of every book that exists in our physical world to exist in the spirit world.

e. When an author writes a book, he or she would first grab the spiritual entity that is the spiritual form of the main topic of the book. This then enables every larger spiritual entity that has that spiritual entity as part of it to be right there next to the spirit of the author.

f. The author would then pick and choose among these spiritual entities and assemble them in a manner that would be the spiritual form of the text he or she wants.

g. However, the assemblage could only be made if it already exists in the spirit world. The author finds the assemblage in the spirit world with his or her spiritual senses, and then translates it into a form that could exist or be expressed in our physical world.

19. **Oneness pervades the spirit world:**

    a. As mentioned earlier, only one copy of any existing piece of knowledge is kept in the spirit world. This mean each of countless spiritual entities that include a given piece of knowledge in their makeup would be sharing at least that given piece of knowledge among themselves as well as possibly other parts of themselves.

    b. This kind of sharing among spiritual entities would be created by every other existing piece of knowledge as well.

    c. A similar kind of sharing among larger spiritual entities would be created by every smaller spiritual entity because larger spiritual entities are made up of smaller spiritual entities. This is because a given smaller spiritual entity would be a part of countless larger spiritual entities.

    This kind sharing is created also because only one copy of any existing piece of is kept in the spirit world.

    d. The net result is everything in the spirit world is directly or indirectly a part of everything else in the spirit world. Therefore, oneness pervades the spirit world, and consequently so does the feeling of love, empathy, and compassion.

    e. This means every human spirit is directly a part of every other human spirit. We know they are directly so, and not just indirectly so, because:

    - We share over 90% of our DNAs among each other.

    - The construction of our physical bodies is the same.

- We all have the same kinds of consciousness, intelligence, senses, instincts, intuition, emotions, abilities, etc., although their levels could vary.

f. Because all human spirits are sharing a sizable part of themselves among one another:

- Ego does not exist in the spirit world.

- No human spirit is better than any other human spirit.

- Only win-win conditions exist. Win-lose and lose-lose conditions are only concepts and are not practices in the spirit world.

20. **The spirit world is physically nowhere in our physical world, but it is spiritually everywhere in our physical world:**

a. The spirit world could not physically exist in our physical world, because it is nonphysical.

b. But it is spiritually everywhere in our physical world, because it is enabling everything that exists in our physical world to exist in our physical world.

21. **The spirit world is in a sense more compact than is our physical world:**

a. For example, the spiritual entity that enables one electron to exist in our physical world also enables every other electron to exist in our physical world. One spiritual entity is more compact than essentially an infinite number of electrons spread out all over our universe.

b. Thus, every electron is identical with every other electron in our physical world, but their states of being would vary. For example, they could be a part of living things, be a part of non-living things, be a part of various forms of energy, be a part of liquid substances, be a part of gaseous substances, etc.

c. The same goes for the spiritual entity that enables all protons to exist in our physical world and for the spiritual entity that enables all neutrons to exist in our physical world.

d. Electrons, protons, and neutrons are among the "building materials" that make up our physical world. The same discussion about electrons, protons, and neutrons also applies to every other building material that makes up our physical world such as subatomic particles, atomic elements, and various kinds of molecules.

22. **Only one copy of anything that is not a "building material" would exist in our physical world:**

a. This explains why no two things are identical in our physical world, aside from "building materials" that make up our physical world.

b. We might ask, what about identical twins, various tiny living things such as insects, for example ants, and manufactured things that seem to be identical?

The answer regarding identical twins is if we look closely, we would see they are not absolutely identical physically or in any other way. The differences could be slight, but that is enough to make then not exactly identical. Also, the experiences each goes through could never be exactly the same. Any of such differences, even if only slight, is enough to make identical twins not absolutely the same.

The same goes for various tiny living things such as insects, for example ants.

c. Regarding manufactured things that seem to be identical, each is actually different if only slightly due to natural variations in the material, colors, etc. Machinery tolerances would cause variations in dimensions, alignments, etc. Consequently, no two manufactured items would be truly exactly the same.

23. **Nothing is perfect, not even the spirit world:**

   a. The spirit world itself is not perfect and could never be perfect because it could never possess every piece of knowledge that could be generated.

   b. New experiences are always possible and thus new pieces of knowledge would always be possible to generate. This is why the spirit world could never possess every piece of knowledge that could be generated.

   c. As stated earlier, a piece of knowledge is analogous to a word. This analogy extends to: there is no end to the pieces of knowledge that could be generated, like there is no end to the new words that could be formed. New words and/or new meanings for existing words continue to be formed as we continue to make technical advancements and new applications of our technical advancements.

   d. It is a vicious circle. New pieces of knowledge generated and added to the spirit world would enable new experiences to be possible, and new experiences would generate new pieces of knowledge.

   e. Because everything that exists in our physical world exists because the spirit world enables it to exist, and the spirit world could never be perfect, nothing in our physical world could ever be perfect either.

   f. We can see this how is the case in our physical world. For example, we could never find anything that is really perfect.

g. However, we would often treasure something because its imperfection makes it unique and very valuable. For example, a coin that is among the very few that have an error on them could be worth a lot more as a collector's item than its face value.

h. Good thing no two things are identical and nothing is perfect. Otherwise every person would be exactly like every other person. Then we would not be able to tell which person is our spouse, and our spouse would not be able to tell which person is us. That would really create a mess. But, on the other hand, who would care. After all, one spouse is just as good as any other, or just as bad as any other.

4.3. A Natural Connection and Numerous Opposing Attributes Exist Between the Spirit World and Our Physical World

The spirit world and our physical world have a natural connection with each other. This connection exists even though the two worlds are mutually opposing regarding their structures and various other attributes.

Because we humans are not fully knowledgeable about life, we are not fully aware of what these two seemingly opposite conditions mean and why it is important for us to become equally knowledgeable about both the spiritual part and the physical part of our life. Our unawareness of these things is one of the reasons we humans have not been able to carry out our lives very wisely and why we have made a mess on Earth.

The discussions in the preceding section, Section 4.2., Items 12 and 13, gave some examples of how it is that we are carrying out our life simultaneously in the spirit world and our physical world. Some major differences between the spirit world and our physical world include the following:

1. All human spirits in the spirit world are a part of each other whereas all human individuals are separate entities in our physical world.

2. The spirit world is nonphysical while our physical world is physical.

3. The spirit world will exist forever (assuming its state of knowledge would be kept reasonably balanced forever) while our physical world will exist for only a while.

The most important factor is that our physical world was designed and brought into being by the spirit world for the primary purpose of helping the spirit world restore reasonable balance in its state of knowledge. This is the main reason our physical world has attributes that are the opposite of those of the spirit world.

In essence our physical world was designed to generate the kinds of new pieces of knowledge that would "pull" the state of knowledge in the spirit world in a direction that is opposite the direction of the imbalance that has developed in the spirit world's state of knowledge. By the spirit world having designed our physical world and brought it into being, a natural connection exists between the spirit world and our physical world.

Mankind is given its high level of intelligence, imagination, and creativity in order to be able to figure out how to fulfill its primary purpose for being here on Earth. We are not supposed to rely on the spirit world to tell us how to get it done. If the spirit world could tell us how to get it done it would already have the pieces of knowledge we are to generate, and it would not have needed to design and bring into being our physical world, our universe.

Regardless of whether we are aware of all this or not, we should certainly be intelligent enough to able to know we were not meant to use our high mental abilities to make the mess on Earth that we have

made. In my mind, what is wrong with us is mainly that the amount of spiritual advancements we have achieved is lagging far behind the amazing amount of technical advancements we have achieved. Thus, we keep applying our technical advancements in extremely bad ways in addition to good and neutral ways.

Chapter Five

Spiritual Qualities that likely Initiated the Spirit World and Gave Life to Living Things

5.1 Spiritual Qualities with Spiritual Forms that Are Not Part of the Spirit World

After the spiritual model in Reference 1 was formulated I started wondering what exists before the spirit world came into existence. If the spirit world has a beginning, which according to the spiritual model in Reference 1, it does, then something has to initiate the existence of the spirit world. Using the spiritual model presented in Reference 1 as the basis, I deduced what that "something" is likely to be. The deduction goes as follows:

1. We human, as well as other living things, have certain spiritual qualities that have spiritual forms that are not spiritual entities. Thus, their spiritual forms are not part of the spirit world. They exist outside of the spirit world, and they do not need the spirit world to enable them to exist.

    Every other physical and nonphysical thing that exists or could be expressed in our physical world has a spiritual form that is a spiritual entity that enables the thing to exist or be expressed, and that spiritual entity would naturally be a part of the spirit world. The spirit world is made of spiritual entities, and all spiritual entities reside in the spirit world as explained by the spiritual model in Reference 1.

    Therefore, there is something very different and unique about these spiritual qualities. They are thus likely to have certain abilities that no other spiritual, physical, or nonphysical thing has, including the spirit world.

These spiritual qualities are:
- **Consciousness**
- **Intelligence**
- **Curiosity**
- **Wisdom.**

2. I say these spiritual qualities do not exist in the form of spiritual entities because they are not things that we could mentally hold like how we could mentally hold a thought, attitude, solution, idea, concept, design, feeling, mood, intuition, etc.

We know how it feels to mentally hold something. For example, when we have a thought, we would feel we are mentally holding something, and that something would be the spiritual entity that is the thought. We use our spiritual senses to find in the spirit world the spiritual entity that is the spiritual form of the thought we want, and we would mentally hold that spiritual entity with our spiritual senses in preparation to express it.

By contrast, we would not get a feeling of mentally holding anything when it comes to our having consciousness, intelligence, curiosity, and wisdom. Also, we do not need to use our spiritual senses to find any of these four spiritual qualities. They are simply there with us always available for us to use.

We might ask: what about when we are curious about something? Doesn't our "being curious about something" be something we could mentally hold? The answer is that "being curious about something" and "having curiosity" are two different things. We could mentally hold our "being curious about something," but we cannot mentally hold our "having curiosity" any more than we can mentally hold our having consciousness, intelligence, or wisdom

The same thing holds regarding "being conscious of something" vs. "having consciousness."

We might also ask: what about imagination and creativity? Wouldn't these be in the same category as consciousness, intelligence, curiosity, and wisdom? I my mind the answer is no. Having imagination and having creativity are having the abilities to imagine and to create. We can mentally hold an ability. For example, we could mentally hold the ability to ride a bicycle, to ski, to play the piano, to write a technical paper, to create a garden, etc. Having imagination and creativity are having abilities much like having the other mentioned abilities.

Having an ability is being able to know how to use our spiritual senses to find something in the spirit world and that something would be the spiritual form of a physical thing such as a technical paper or a garden or it could be the spiritual form of a nonphysical thing such as how to ride a bicycle or how to ski or how to play a piano.

Therefore, having imagination and having creativity involve having spiritual senses and the ability to use the spiritual senses in the manner described. Accordingly, they are not in the same category as consciousness, intelligence, curiosity, and wisdom.

It is for such reasons that I conclude the spiritual forms of the four spiritual qualities: consciousness, intelligence, curiosity, and wisdom are not spiritual entities but are something else. In fact, their spiritual forms are not the things that enable them to exist; their spiritual forms are simply what these four spiritual qualities are. Therefore, we need not talk about their spiritual forms; we can simply talk about them.

Thus, we could simply say the following:

The four spiritual qualities: consciousness, intelligence, curiosity, and wisdom are not spiritual entities. Therefore, they do not reside in the spirit world. They reside outside the spirit world.

3. We might ask: if consciousness, intelligence, curiosity, and wisdom are not a part of the spirit world, they could not be a part of our spirit, which is a part of the spirit world, and therefore they are not a part of us. So, how could we say we have these spiritual qualities when they could not be a part of our spirit? For simplicity, let's denote these four spiritual qualities with the symbol "4Qs."

The answer is we are constantly connected with the 4Qs. Thus, we could call these connections as being connections of the third kind. How this works is explained as the presentation continues.

Recall that Reference 1 introduced the idea of connections of the first kind that connects every exiting piece of knowledge with every other existing piece of knowledge, and connections of the second kind connects every piece of knowledge making up a spirit of a living or nonliving thing with the physical forms of the living or nonliving thing that resides in our universe. All the connections of the second kind combined would form the soul of the living or nonliving thing.

5.2. The 4Qs Likely Initiated the Existence of Knowledge and the Existence of the Spirit World

So far, we are likely to think that before the spirit world came into existence that there is simply nothingness. But in order for the existence of the spirit world to begin, something has to initiate its existence. Therefore, there has to be something that exists before the spirit world came into existence. Since the 4Qs are not part of the spirit world, they could exist before the spirit world came into existence. And, if the 4Qs are the only things that exist before the spirit world came into existence, they are then likely to be what initiated the existence of the spirit world. In order for them to do that, they must be what initiated the existence of knowledge and the existence of life for living things.

Notice that the spiritual qualities of consciousness, intelligence, curiosity, and wisdom are the spiritual qualities of living things. Therefore, it might be that these spiritual qualities could give life to living things.

In order for the 4Qs to generate new pieces of knowledge they have to go through experiences. The kinds of experience they could conceivably go through would include the following:

a. Experimentations.

b. Questionings.

c. Explorations.

d. Searches, including searches for states of balance.

e. Awareness.

However, the nature of such experiences they could go through would be uniquely different from the nature of such experience a normal living thing would go through because the 4Qs are not living things in the same sense that a normal living thing would be a living thing. Therefore, the new pieces of knowledge they would generate would also be different in nature from new pieces of knowledge a normal living thing would generate.

It is possible that the nature of the new pieces of knowledge they would generate could initiate the existence of life of the nature we are used to perceive as being life.

It is also possible that by generating enough of the right new pieces of knowledge that the 4Qs could initiate the existence of the spirit world as a living thing. Each of those new pieces of knowledge would have a connection of the third kind connecting it to the 4Qs.

The combination of all these connections of the third kind would form the soul of the spirit world, and it would be an ultra-soul. The 4Qs would then be the spirit of the spirit world, and it would be an ultra-spirit.

This would be similar to how a thing existing in our physical world would have a spirit in the spirit world enabling it to exist, and it would have a soul consisting of all the connections of the second kind connecting it with its spirit.

In the case of the spirit and soul of the spirit world, since the 4Qs reside in a spiritual location that is beyond the spirit world, we could say the 4Qs are ultra-spiritual and that the spirit and soul of the spirit world are an ultra-spirit and an ultra-soul simply to denote that the 4Qs are spiritually beyond what we would perceive as being spiritual. Accordingly, then, the pieces of knowledge generated by the 4Qs could be considered ultra-pieces of knowledge just to distinguish them from the usual pieces of knowledge.

5.3. Because the Spirit World Is a Living Thing, Almost Everything It Creates Is a Living Thing; Some Being More Living than Others

The ultra-pieces of knowledge are thus a part of the spirit world just as any usual pieces of knowledge are a part of the spirit world. The difference is the ultra-pieces knowledge together with the ultra-spirit of the spirit world give life to the spirit world. Just as any piece of knowledge in the spirit world, the ultra-pieces of knowledge would form connections of the first kind with every other piece of knowledge in the spirit world whether the other pieces are the usual kind or are the ultra kind.

Thus, the ultra-pieces become part of the oneness that pervades the spirit world, and almost every spiritual entity would have one or more ultra-piece of knowledge as part of it. This means almost every spirit the spirit world creates would have one or more connection of the third kind with the 4Qs such that almost everything the spirit world creates would be a living thing. Some being more living than others depending on the number of connections of the third kind their spirits have and how complexity of their spirits. As explained in Chapter Eight, gradations and diversities are natural and very important attributes of the spirit world.

This is consistent with the explanation presented in Reference 1 describing how it is that everything that exists in our physical world has some degree of consciousness and intelligence, even if it is on an elemental level. An example of elemental consciousness and elemental intelligence would be how water could be perceived as being conscious of its temperature and would know when to freeze, or melt, or evaporate, etc. Another example would various chemicals be perceived as being conscious of the condition that they are in and would know when to react and when not to react.

This could also explain how it is viruses and prions could be perceived as being somewhere between being living things and nonliving things.

The level in which a spiritual entity would have of each of the four spiritual qualities would largely depend on the evolutionary history of that spiritual entity. The more varied the history. the more frequent the experiences the spiritual entity have gone through, and the longer the history the higher would be the levels of the spiritual qualities the spiritual entity would have.

From our observations of people on Earth, there are different kinds of consciousness, intelligence, curiosity, and wisdom, and each individual would have a different combination of kinds regard these four qualities. This would explain why different individuals are good at doing different things. Some would make good scientists, some would make good business people, some would be good in sports, some would be good at repairing things, etc.

Of the four spiritual qualities, wisdom is the only one that has special requirements in order to be able to be put into action. It is that the state of knowledge of a living thing needs to be reasonably balanced. Then the living thing would be able put that spiritual quality into action to form good wisdom that is relevant to the knowledge involved.

5.4. The 4Qs Initiated the Existence of the Spirit World as a Living Thing for a Reason

A likely reason the 4Qs initiated the existence of a living spirit world could be as follows:

1. Having consciousness, intelligence, curiosity, and wisdom and having knowledge are two different states of being, although we humans often confuse one with the other. While the 4Qs are limitless in the levels of their qualities, they do not have any knowledge on their own by themselves.

2. Therefore, a likely reason the 4Qs initiated the existence of the spirit world as a living thing is because spirit world is limitless in the amount of knowledge it could generate and possess. But on the other hand, the spirit world does not have the qualities of the 4Qs on its own by itself such that it could figure out how to be viable and stay viable.

3. Therefore, the 4Qs and the spirit world together make a great team that could do essentially anything and everything that neither one could do on its own.

This is likely why the 4Qs initiated the existence of the spirit world as a living thing.

5.5. The 4Qs Most Likely Initiated the Existence of the Spirit World Multiple Times

In my opinion there are likely to be a lot of parallels between how things work in our physical world and how things work in the spirit world and beyond. The "beyond" pertains to the 4Qs in that I am assuming they exist before the spirit world came into existence. One of the parallels is that nothing is perfect or could be perfect. Therefore, during the period in which the 4Qs are working on initiating the existence of the spirit world as a living thing, something like the following most likely happened.

1. All living things on Earth begin life with a very vulnerable early period. A lot would not survive during this early period. Possible reasons include; environmental conditions, insufficient food, predators, something deficient in the make up of the living thing, etc.

2. When the 4Qs generate enough ultra-pieces of knowledge that could form a spirit world, it is likely that they had to do this multiple times before they come up with a version of a spirit world that is able to survive its vulnerable early period.

3. Each version would be made of a different combination of ultra-pieces of knowledge. The 4Qs are able to figure out what combinations might work, but because the 4Qs could not retain knowledge and therefore do not have a memory, each version is essentially a brand-new experiment starting from scratch. This is the reason why I think the 4Qs had to initiate the existence of a spirit world multiple times before finally coming up with a version that was able to survive its vulnerable early period.

I also believe this is the source of why it is that all living things on Earth are designed to begin life with a vulnerable period. It is because this is what the spirit world has gone through and has found it to be workable even if not perfect.

4. We might ask: is it possible that more than one version of a living spirit world managed to survive its vulnerable early period? The answer is yes. But if more than one version were to survive their early periods, they would merge and only one will remain. After all, both versions would be doing the same thing and thus serving the same purpose. Anything spiritual would likely go for the simplest way to do things, and therefore one spirit world would be better than two.

5. A more realistic situation would be that if one version of a spirit world is to survive its vulnerable early stage, the 4Qs is not likely to try initiating a second version.

5.6. The Size and Complexity of a Spiritual Entity Do Not Matter When It Comes to Being a Living Thing or Not

We might think the larger and more complex spiritual entities would more likely be living things. However, this is not necessarily the case. For example:

1. The spiritual entity that enables a giant airliner to exist on Earth would be large and complex, but it is not a living thing.

2. On the other hand, the spiritual entity that enables a tiny ant to exist on Earth would be much smaller and simpler, and it would be a living thing.

5.7. A Possible Source of Why a Male Component and a Female Component Are Needed to Produce Living Things

The 4Qs and the spirit world need each other in order to produce living things. Each provides something that the other cannot, and together they are able to provide what is needed to produce living things.

This could be the source of why a lot of living things on Earth need a combination of a male thing and a female thing to produce living things. Each provides something that the other cannot, and together they are able to provide what is needed to produce living things.

Again, I think the spirit world would use a process or procedure that is patterned after something it has experienced and knows would work.

5.8. Living Things Are Most Certainly to Exist on Other Hospitable Planets Besides on Earth

The conventional thinking is that certain life producing molecules came to Earth on meteors, and such molecules eventually formed very primitive living things such as slimes. These primitive living things eventually evolved to become the currently existing living things on Earth.

If what is presented so far about the 4Qs and the spirit world is true, then living things are bound to exist on other hospitable planets in our universe beside on Earth. The mechanism for how life could emerge on Earth and on other hospitable planets would be due to the work of the 4Qs and spirit world instead of by life producing molecules arriving on Earth and other planets carried on meteors.

5.9. A Spiritual Place Exists Beyond the Spirit World That Is the Origin of Life

Up to now our human imagination has been mostly able to reach out and touch a spiritual place that is one step removed from our physical world, and we call this place "heaven", "the spirit world", "the other side", etc. What if there are additional spiritual places that are beyond what we have been referring to as heaven, the spirit world, the other side, etc.?

I think in order to speculate how the existence of the spirit world got initiated, we have to allow for spiritual places that are one or more steps removed from the spirit world. Such a spiritual place was envisioned as where the 4Qs reside. So far, by adding the concept of the existence of the 4Qs to what have been envisioned in Reference 1, plausible explanations have been possible for additional everyday common observations and experience. Examples include:

1. How living things get their consciousness, intelligence, curiosity and wisdom qualities.

2. Why some living things have more of such qualities than other living things.

3. How it is possible for things such as viruses and prions could be in between being living things and nonliving things

4. How it is possible for some things to have elemental consciousness and elemental intelligence.

5. How it is that every living thing on Earth begin life with a very vulnerable early stage.

6. Possible explains how it is that almost every living thing on Earth needs a combination of a male living thing and a female living thing to produce new living things.

A plausible concept is that something exists that is the origin of life. The title of this book is *Understanding the Spirit World and Beyond*. The "Beyond" indicates there is a place "beyond" the spirit world in which everything there is likely to be living. It is the place that is the origin of life, and it is the place that has the power to give life to the spirit world and to the universes that are brought into being by the spirit world.

Chapter Six

The Three Major Obstacles That Are a Part of Our Physical World

6.1. Our Physical World Was Designed to Be Opposite the Spirit World

The next chapter in this book, Chapter Seven, explains mankind's primary purpose for being here on Earth. In order for mankind to be able to go through the kinds of experiences that would generate the kinds of new pieces of knowledge that would fulfill mankind's primary purpose, mankind has to have a world that would provide the opportunities for going through such kinds of experiences. As explained in Chapter Seven this means the world mankind needs to reside in has to have features that are opposite those of the spirit world.

Our physical world was designed and brought into being by the spirit world to meet this need. For example:

1. The spirit world is nonphysical. Our physical world is physical.

2. Human spirits are a part of one another in the spirit world. Human individuals are separate entities in our physical world.

3. Lifespan lasts forever in the spirit world. Lifespan lasts only for a while in our physical world.

4. The spirit world is unlimited in dimensionality. Our physical world is limited to three-dimensions.

5. The spirit world is made of a variety of different pieces of knowledge, none of which are duplicates. Our physical world is made of the same variety of materials throughout.

Because of such necessary differences between the spirit world and our physical world, various obstacles that were not purposely designed into our physical world ended up being a part of our physical world. This created some unintended traps mankind could fall into as explained in the next section of this chapter.

A more extensive list of opposites between the spirit world and specifically Earth is given in next chapter in this book, Chapter Seven "Mankind's Primary Purpose for Being Here on Earth".

6.2. Spirituality and Science Need to Be Pursued Together, not Separately

Among the obstacles that ended up being a part of our physical world, the three major ones are:

Obstacle (1): Our universe is made of materials and energies that can be possessed, controlled, and/or consumed.

Obstacle (2): Highly intelligent living things such as us humans on Earth have egos and a desire to enhance survivability.

Obstacle (3): Mobile intelligent living things (those that can walk, swim, and/or fly) need to eat other living things to survive.

These obstacles have the following opposite features in the spirit world:

- The spirits of living things do not possess, control, or consume things and energies in the spirit world.

- The notions of ego and desire to enhance viability are only concepts in the spirit world, not practices.

- Living things do not need to eat anything in the spirit world to stay viable.

Since these obstacles were not purposely designed into our physical world, the intent was for mankind to figure out ways to work around them and not be trapped by them. The spirit world must have figured that since mankind is highly intelligent, imaginative, and creative that it would be able to figure out how to work around them. However, the chance of getting trapped by them was there, and unfortunately mankind did get trapped by the obstacles.

In my mind, a reason mankind got trapped is because mankind pursued spirituality and science separately. Mankind did not take into consideration spirituality in its pursuit of technical advancements and thus did not do anything to help prevent bad applications of its technical advancements. A lot of the bad applications have to do with being trapped by the three major obstacles.

This makes the effort mankind must put in to fulfill its primary purpose much more difficult, because now it has to first find a way to get out of the trap. Up to now mankind has not found a way, or perhaps mankind doesn't realize it got caught in a trap. Because mankind has been in the trap for such a long time (for thousands of years) it thinks its current way of life on Earth is normal. In my opinion it should not be considered normal. The potential is there to make it much better. According to the spiritual model presented in Reference 1, making it much better is part of mankind's primary purpose for being here on Earth.

Consequently, mankind has invested so much in support of its current way of life that anyone who might threaten to change it would be persecuted one way or another. Even the various efforts attempting to get mankind out of the trap, such as the various religions, are not free from having some degree of investment in mankind's current way of life.

As a result, their efforts have declined in effectiveness as evident by declining church attendance especially in today's high-tech world in which mankind's technical advancements have surged way ahead of mankind's stalled spiritual advancements.

When mankind's technical advancements are surging way ahead of its spiritual advancements, and worse when its spiritual advancements are stalled, bad applications would overtake good applications. This makes life ever increasingly complicated which in turn increases frustration, impatience, and disenchantment regarding just about everything. This gets mankind deeper into the trap. Mankind needs to find a way to greatly reduce and eventually eliminate its desire to do bad applications of its technical advancements.

This would go a long way toward improving how mankind carries out its lives. It would also shorten and eventually eliminate the time it takes to deal with all the bad applications of its technical advancements thus leaving more time and effort to work on improving how it carries out its lives.

In my opinion, mankind needs a new way of thinking about what life is about. In relation with the current way of thinking, the new way would have to be unconventional. One possible and more a realistic way is to understand more thoroughly that life is equally spiritual and technical. Right now, science and spirituality are perceived to be separate such that one is seen as having nothing to do with the other.

I think this separation has a lot to do with mankind allowing itself to give in to the trap it got caught in and thus think it is worth pursuing bad applications of its technical advancements as if it is some kind of a basic right to do so. After all, that is how some of the super rich got to be super rich.

Thus, I think keeping spirituality and science separate is incorrect, counter productive, and bad for mankind's evolutionary progression. The spirit world designed and brought into being all that our physical world is made up spiritually as well as physically. Therefore, spirituality and science belong together. Thus, they need to be pursued together, not separately.

Chapter Seven

How to Fulfill Mankind's Primary Purpose for Being Here on Earth

7.1. The Spirit World Needs to Relearn How to Restore Balance Each Time an Imbalance Develops

s stated in Chapter One, we humans have a primary purpose and a secondary purpose for being here on Earth.

1. The primary purpose is to help restore balance in the state of knowledge of the spirit world.

2. The secondary purpose is to generate as many new pieces of knowledge as possible to help the spirit world continue to grow.

The spirit world's state of knowledge would naturally develop an imbalance periodically, and it would then need to learn once more how to restore balance each time that happens.

The reason the spirit world needs to learn how to do the rebalancing all over again each time is because the spirit world grows and is different each time. It would usually design and bring into being a new universe to help do the rebalancing. Each new universe would be unique and different from any that came before, because the spirit world has grown and is different each time such that what is needed to do the rebalancing would be unique and different each time.

An imbalance would develop periodically because the spirit world has lots of ongoing tasks associated with managing its growth and keeping its state of knowledge balanced such that it is able to form the wisdom needed to figure out how to stay viable. This means the spirit world is constantly going through experiences that take place within the spirit world itself. But there are numerous things that are only concepts and are not practices in the spirit world such as; kill, insult, greed, hate, etc.

The spirit world deals with such things only as concepts and not as practices. In addition, much of the experiences it goes through are ones that could be formed only within the spirit world itself. Consequently, the spirit world eventually becomes overloaded with new pieces of knowledge to do with concepts and to do with experiences that could take place only within the spirit world.

Therefore, the spirit world has two major causes of imbalance. It needs then to have new pieces of knowledge to do with practices of such things that it has been dealing with only as concepts, and it needs also to have new pieces of knowledge generated from experiences other than those that could only be formed within the spirit world itself. This explains why the spirit world would periodically design and bring into existence universes that could fulfill both needs. Universes could be designed to address both of the major causes of imbalance.

We might ask; could we explain more clearly why growth would cause the spirit world to has to learn all over again how to do rebalancing each time rebalancing is needed?

As the spirit world grows, new things are created and existing things evolve such that the spirit world of today is very different from how the spirit world was at any time in the past. Therefore, whatever was learned about restoring balance in the past would no longer be relevant for the present.

By analogy, today's world on Earth is very different from how the world was at any time in the past. Thus, any approach used to resolve issues of the past are bound to need updating to be relevant for the present, and usually completely new approaches are need to address completely new issues.

An example would be the issues that have developed regarding bad applications of social media. Any approach used to resolve issues of the past before social media came into being would have to be updated to work on the issues to do with social media, and more likely completely new approach might be needed instead.

A similar thing happens in the spirit world, and that is why when an imbalance develops, the spirit world needs to learn anew how to restore balance. Accordingly, each time the spirit world would design and bring into existence a new universe to help restore balance, the new universe will be unique and different from any of the past.

7.2. A Certain Requirement Must Be Met for the Restoration of Balance to Work

As explained in Reference 1 and also in Chapter One, because only one copy of every existing piece of knowledge is kept, love, empathy, and compassion, would naturally pervade the spirit world. This attribute will thus always remain in place no matter what new pieces of knowledge are generated and added to the spirit world.

Therefore, this attribute will also be a part of any imbalance that develops. This means any new pieces of knowledge that are generated for the purpose of restoring balance must also have this attribute as part of the experiences from which they are generated in order for the restoration process to work.

In other words, the attribute of love, empathy, and compassion must be a part of the activities and therefore of the experiences that generate the new pieces of knowledge in order for the new pieces of knowledge to be "appropriately aligned" with those pieces of

knowledge that are causing the imbalance such that the new pieces of knowledge are able to restore balance.

Therefore, a summary statement of mankind's primary purpose for being here on Earth would be as follows:

Mankind's primary purpose for being here on Earth is to gain as much knowledge as possible about our physical world and the spirit world, and to apply that knowledge to make mankind's life on Earth to resemble as close as possible to how life is in the spirit world.

Otherwise, the new pieces of knowledge would only go toward fulfilling mankind's secondary purpose.

By analogy, the attribute of people to people interactions pervades our physical world. This attribute of our physical world could be perceived as being analogous to the attribute of love, empathy, and compassion that pervades the spirit world. Issues in our physical world that need resolving almost always involve people to people interaction. Therefore, any procedure to be used to resolve any issue to do with people to people interaction would have to have the attribute of people to people interaction as part of it in order for the procedure to work.

What this means by analogy is that in order for mankind to be able to fulfill its primary purpose for being here on Earth, it has to change its current state of human nature to one that would have mankind always behaving with love, empathy, and compassion.

Given that mankind's current state of human nature is far from being this way, it will take numerous generations for such a change to occur. But we need to start somewhere.

To help mankind make this change, I firmly believe we need to consistently help young people and children develop critical thinking skills. The future of mankind is in always in the hands of young people and children. From my observations of human behavior, I think a major reason mankind made a mess on Earth is because we humans on the average are insufficient in our critical thinking skills, as well as being insufficient in our spiritual advancements.

7.3. What Does Mankind Need in Order to Fulfill Its Primary Purpose?

Mankind's primary purpose is to help restore balance in the state of knowledge of the spirit world.

1. More specifically, mankind's primary purpose for being here on Earth is basically to generate the kind of new pieces of knowledge that would help restore reasonable balance in the spirit world's state of knowledge.

2. The preceding section, Section 7.2., pretty much explains what mankind needs to do to fulfill its primary purpose. But, given mankind's current state of human nature and how mankind has been behaving overall poorly for thousands of years, mankind most likely needs to be more inspired and more motivated to do what is described in Section 7.2.

3. What has been missing so far is a way to better understand what the spirit world is made of and how things work in it. I think the absence of such information is a major reason mankind lost its way as to how to fulfill its primary purpose.

4. And, after thousands of years, mankind essentially forgot it has a primary purpose for being here on Earth, although subconsciously most individuals most likely believe mankind has a purpose but have not been able to figure out or recall what it is.

This book and Reference 1 are written to provide mankind with a more complete understanding of the spirit world and a new and more effective way to pursue spiritual advancements. This would hopefully inspire and motivate mankind from within itself to pursue spiritual advancements and to resume making progress toward fulfilling its primary purpose.

7.4. How Mankind's Secondary Purpose Will Be Fulfilled

Mankind's secondary purpose is to generate as many new pieces of knowledge as possible to help the spirit world continue to grow. This would also help assure mankind would go through as many as possible all the experiences that could generate the kind of new pieces of knowledge that restores balance in the spirit world.

How mankind's secondary purpose could be fulfilled is described by the following statements:

1. Mankind's secondary purpose would be fulfilled essentially automatically since each of us humans would be more or less constantly going through experiences while awake and thus would be constantly generating new pieces of knowledge while awake.

2. Mankind is given the freedom to choose such that it would not be restricted as to the kinds of new pieces of knowledge it would generate. This is so that mankind would generate as many as possible new pieces of knowledge of any kind and thus help the spirit world grow as much as possible and in as many ways as possible.

   This also helps to increase diversity within the spirit world. Diversity is one of the major sources of strength for the spirit world as explained in Chapter Nine.

3. The risk the spirit world takes is that mankind could become so interested in simply generating new pieces of knowledge that it

might not focus on fulfilling its primary purpose for being here on Earth. And, sure enough, this has happened.

4. The spirit world is willing to take this risk because if it restricts mankind to generate only the kind of new pieces of knowledge that would restore balance in the spirit world it would in effect be doing the thinking for mankind.

   If the spirit world knows how to do the thinking for mankind for restoring balance in the spirit world, it would already have the pieces of knowledge it needs to restore balance, and therefore it would not have needed to design and bring our universe into being.

5. Therefore, the spirit world decided to trust mankind to eventually do what is needed to fulfill its primary purpose of helping the spirit world restore balance while at the same time helping the spirit world continue to grow.

7.5. A Manager of a Complex World Needs Help to Balanced its State of Knowledge

We might ask; why is it important that the spirit world designs and brings into being universes to help restore balance in its state of knowledge? Why simply having some things as concepts and not practices not be enough to maintain balance?

The answer is basically that the experience of working with something only as a concept is different from the experience of working with it as a practice. The pieces of knowledge that would be generated would be different. When the spirit world has generated a lot of pieces of knowledge that are associated with concepts, sooner or later it will need to counterbalance them with pieces of knowledge that are associated with practices of the same things.

We might then ask; why are pieces of knowledge associated with concepts and pieces of knowledge associated with practices are mutually counterbalancing? The answer is as follows:

1. When we are dealing with something by practicing it, we are going to learn more completely about that something than if we are only dealing with it as a concept. It is like we could learn how to ski much better by practicing skiing than by simply reading about how to ski.

2. Having balanced knowledge about something means having a more complete knowledge about something. Practicing something would provide a more complete knowledge about something than to just deal with it as a concept. Therefore, it is in this sense that the knowledge generated by practicing something would create or restore balance for the knowledge that was generated by simply dealing with something as a concept.

3. Anything could be dealt with as a concept in any world. But not everything could be practiced in the spirit world because it is filled with love, empathy, and compassion, and it is a nonphysical world.

4. For example if the thing was "kill," the concept of killing could be dealt with as a concept in the spirit world but not as a practice. This is because nothing could be killed in the spirit world. In addition, the act of killing is not compatible with the love, empathy, and compassion that fill the spirit world.

    Therefore, the spirit world has to design and bring into being a world such as our physical world in which killing could be a practice in order to generate pieces of knowledge associated with the practice of killing.

5. For another example if the thing was "mistreat." The concept of mistreating could be dealt with in the spirit world as a concept

but not as a practice. This is because mistreating anything would not be compatible with the love, empathy, and compassion that fill the spirit world.

In this case the spirit world could design and bring into being either a physical or a nonphysical world into being in which mistreating could be a practice. Such a world would not be filled with love, empathy, and compassion.

This implies that our physical world is not filled with love, empathy, and compassion since we humans do kill and mistreat other humans. So, does that mean mankind on Earth could never fulfill its primary purpose?

I say the answer is mankind can fulfill its primary purpose. It just means our physical world is not automatically filled with love, empathy, and compassion. That is why living things need to kill other living things in order to survive as discussed in Chapter Six. But it doesn't mean we could not make mankind's life on Earth to be infused with love, empathy, and compassion such that it would be as close as possible to being like how life is in the spirit world.

This discussion has some important implications regarding how a manager of a complex world could go able managing with a lot of wisdom. The spirit world is a prime subject for this discussion.

The spirit world is its own manager of itself, and the spirit world is as far as we know the most complex world that exists. As pointed out in Reference 1 and also in this book, the spirit world is always incomplete and imperfect. This is because the spirit world could never possess every piece of knowledge that could be generated.

However, the spirit world could always achieve balance in its state of knowledge. Having balance enables it to form wisdom, and wisdom is vitally important to enable the spirit world to figure out how to keep it self viable and how best to manage it self and its own growth.

By analogy, the president of a very complex nation is one of several managers of that nation. No one could possible have practiced everything that goes on in a complex nation. For example, the person who becomes president tends to come from a wealthy background more often than from a middle class or poor background. Therefore, chances are that person would have in his or her mind only concepts and not practices of how it is to be middle class or poor. And yet he or she is to manage a nation made up mostly of middle class and poor people.

A good manager would thus include among his or her advisors, people who have practiced things that the manager has only concepts about and that are important for the wellbeing of the nation. Recruiting such people would be analogous the spirit world designing and bringing into being universes to practice the things the spirit world has only concepts about and that are important for the well being of the spirit world. In the case of a manager of a nation, having a balanced state of knowledge for his or her administration could enable the manager to manage with a good measure of wisdom.

What this indicates is if a manager includes in his or her choice of advisors only people with the same background as his or her background, he or she would be managing without the wisdom necessary to manage well. This is especially the case if the manager and advisors all came from wealthy backgrounds and the population consists mostly of middle class and poor people.

While some very wealthy people have had practices being middle class or poor either directly personally or indirectly through friends and relatives, most very wealthy people are not likely to have had such practices.

What all this means is during election periods, ideally a lot of details about the backgrounds of candidates ought to be revealed. And the candidates need to be asked who they would recruit as their advisors. The background of all such people ought to also be revealed. This will indicate if a candidate will govern with a lot of wisdom or with very little wisdom. This could also indicate whose interest the candidate has in mind, the nation's or his or her own. To my knowledge such an assessment has never been done or has been used as an approach to decide who to vote for, but it should be used.

7.6. What Could Be Involved in Restoring Balance Besides Mankind's Primary Purpose

Any living thing that lives for a long time, and therefore grows for a long time, would need periodic adjustments to stay in good shape. Various species of trees such as fruit trees need regular pruning, corporations and various other businesses need to frequently adjust their business plans to keep up with changing times, fast growing cities and communities need to keep making changes to accommodate their growths, etc. The spirit world similarly needs to regularly restore balance in its state of knowledge to remain viable.

If the spirit world were to grow for an extended period with new pieces of knowledge generated mainly from one source, its state of knowledge would eventually develop an imbalance, because it would have too many pieces of knowledge coming from that one source. The spirit world would then design and bring into being a new source of new pieces of knowledge that could restore balance. The new source is likely to be a new universe.

The spirit world always has lots of things going on such that it is bound to want the new source to do other things as well as restore balance. Therefore, the new source would be designed to enable other kinds of experiences to take place besides the kind that would generate the kind of new pieces of knowledge that could restore balance.

Sometimes it could require multiple simultaneously existing sources to accomplish everything the spirit world has in mind. In our case, this means there might be other universes existing besides ours. This could explain why we have UFOs and UAPs visiting Earth. They might be carrying visitors from other universes.

The multiple simultaneous sources could be different from each other. One source might be physical and dimensionally different from the others, and another source might be nonphysical, etc. Sometimes it could require a sequence of sources, one acting after another after another, etc. This is analogous to a person transferring from one transportation mode to another and another to finally reach his or her destination.

On occasion, restoring balance in the spirit world might happen to be especially tricky such that it requires a sequence of small adjustments, each of a different nature, instead of one large adjustment of a single nature. It would be interesting to know if our universe were meant to make one large adjustment for a complete rebalancing or is it meant to make a small incremental adjustment.

In any case, Earth seems purposely designed to generate the kind of new pieces of knowledge that counterbalances the kind of pieces of knowledge that is generated by experiences that could take place only within the spirit world itself. An indication of this is that the attributes of Earth are generally opposite those in the spirit world.

According to the spiritual model presented in Reference 1 the attributes of the spirit world are the following:

1. The spirit world is nonphysical.

2. Everything is directly or indirectly a part of everything else. Therefore, every living thing is a part of every other living thing.

3. The spirit world is unlimited in dimensionality.

4. Communications are instantaneous throughout the spirit world and are always accurate.

5. A leader is not needed to bring every living thing together. They

simply do everything together since they are all directly or indirectly a part of each other. Therefore, no living thing would feel insecure, alone, distrust, or fear, and no living thing would have the urge to fight or flight.

6. Competition, having an opponent, and win-lose and lose-lose situations are concepts, not practices.

7. Nothing could be solely possessed or controlled by any one living thing.

8. Everything exists forever; nothing ever dies as long the state of knowledge in the spirit world is kept balanced.

9. The spirit world is made of a variety of different pieces of knowledge, none of which are duplicates.

10. Ego, envy, jealousy anger, fear, killing, battles, hate, evil, etc. are only concepts and are not practices. Some of such things are feelings, and most of such feelings are only concepts and are not practices.

The attributes of Earth are the following: They are generally opposite those of the spirit world:

1. Our physical world is physical so that Earth is physical.

2. All living and nonliving things are separate physical entities.

3. Our physical world is three-dimensions and so is Earth.

4. Communications are not instantaneous and are sometimes inaccurate.

5. A leader is needed to bring everyone together, although not all leaders could or would do that.

6. Competition and having an opponent to compete or fight against are a part of almost everything we do. Win-lose and lose-lose situations are more often the outcome instead of win-win situations.

7. Things and energy could be possessed and/or controlled by individuals.

8. Living things do not live forever.

9. Our physical world is made of the same variety of materials throughout and so is Earth.

10. Ego, envy, jealousy, anger, fear, killings, battles, hate, evil, etc. exist on Earth.

Thus, we can see how it is that our physical world was designed to naturally generate the kind of new pieces of knowledge that could counterbalance the kind of new pieces of knowledge the spirit world would generate by going through experiences that exist only within the spirit world it self.

One more requirement the new pieces of knowledge we humans generate on Earth must meet in order for them to go toward fulfilling mankind's primary purpose. The experiences that generated those new pieces of knowledge must lead to making mankind's life on Earth to be as close as possible to being how life is in the spirit world. This is explained in Section 7.2.

7.7. Generating New Pieces of Knowledge that Restore Balance that Are Also Compatible with How Life Is in the Spirit World

The follow are some examples of things that are only concepts and not practices in the spirit world, and that are among the things our physical world was designed and brought into being to address.

Cheat
Competition
Destroy
Discriminate
Disease & Illness
Dishonesty & lies
Ego
Envy
Greed
Harm
Hate
Insult
Kill
Mistreat
Opponent
Poison
Steal & rob
Threat
Wars & Battles
Win-lose
Etc.

We humans are to address things such as these in a manner that would lead to mankind's life on Earth to be as close as possible to being how life is in the spirit world. This means we are to assume the attitude of love, empathy, and compassion while going through the experiences that would generate the new pieces of knowledge regarding such things. It can be done, but it takes a lot of creativity, and some things are more easily done than others.

The following are some examples of things that are OK to do:

1. In the case of "kill," it's OK to kill only for food and for self preservation.

2. In the case of "steal and rob," it's OK to steal eggs for food from a chicken's nest and steal milk for food from a cow, etc.

3. In the case of "envy," I suppose if we simply envy something but do not do anything bad in response. That might be OK.

4. In the case of "greed," perhaps being greedy to some limited extend but then are willing to donate what we are greedy about to help others who need help might be OK.

5. In the case of "discrimination," if we discriminate only in terms of who is capable of doing a job and who are not capable of doing a job to determine who would get the job, then that should be OK.

I picked five somewhat easy cases. The rest are not likely to be as easy, and there are a lot more that are not included in the list.

Fulfilling mankind's primary purpose is not easy. If it were easy, I would imagine the spirit world would have done something simpler than to design and bring into being sometime as complicated as our physical universe to fulfill the purpose. This is likely also why the spirit world selected mankind with its high level of consciousness, intelligence, creativity, etc. to fulfill the purpose. Therefore, we humans better hurry up and get our act together.

7.8. Of Particular Interest Is Why Evil in General Exists on Earth

The reason why evil in general exists on Earth is of particular interest is because of the following:

1. A lot of individuals have asked; if God is a loving and good God, why he or she would allow evil to exist on Earth.

2. According to the spiritual model presented in Reference 1, any perception of God would only be a portion of the spirit world and definitely not the entire spirit world. This is because the spiritual form of everything that exists in our physical world, or in any other possible existing universes, would reside in the spirit world, and each could be only a portion of the spirit world just as the spiritual form of any perception of God could be only a portion of the spirit world.

   This means no one thing in the spirit world, such as a perception of God, could control what the spirit world has to deal with. Evil is simple among the things the spirit world has to deal with just as a lot of other things such as crime, hate, rob, etc. Things such as these could only be dealt with as concepts in the spirit world because they are incompatible with the love, empathy, and compassion that pervade the spirit world.

   But in order to restore balance, such things have to be somehow dealt with as practices. Our physical world happens to be a universe in which evil could be a practice. Therefore, evil was selected along with a lot of other things to exist on Earth for this reason.

3. The spirit world has an imbalance in its knowledge regarding evil because evil is dealt with only as a concept and not as a practice in the spirit world.

   Therefore, spirit world designed our particular physical universe such that evil and numerous other things could be a practice in it, and we humans are to generate the kinds of new pieces of knowledge that would help the spirit world become balanced in its knowledge regarding evil and the numerous other things.

4. Unfortunately, mankind got trapped by the major obstacles that are a natural part of our particular physical world and has gone way too far with its evil practices.

5. Mankind was meant to learn about evil and other things and to apply the knowledge, along with all the other knowledge it learns about our physical universe and the spirit world, in a manner that would make mankind's life on Earth be as close as possible to be how life is in the spirit world.

6. In other words, mankind was not meant to practice evil in all the bad ways it has done. Mankind was meant to perhaps practice evil only mildly without hurting anyone.

   It is also possible that mankind was meant to imagine evil practices but were not to actually carry them out. Because mankind lives in a physical world, it would be able to imagine doing things in a physical world in a manner that the spirit world could not do because the spirit world is nonphysical.

   Therefore, it is this manner that the spirit world intended mankind to generate the kinds of new pieces of knowledge that would help restore balance in the spirit world's state of knowledge.

7. Unfortunately, now that mankind is trapped by the major obstacles and has thus gone way too far in its practices of evil, fulfilling its primary purpose for being here on Earth will be much more difficult.

8. Even worse is that mankind has been behaving so poorly for such a long time, it tends to think the current way life is on Earth is simply the way life is and will always be. It has essentially forgotten it has a primary purpose for being here on Earth.

9. Therefore, what mankind now needs is something to remind it of its primary purpose. Because mankind has become set in its ways, that certain something has to be unconventional. And being unconventional is going to be met with a lot of resistance from individuals who have a lot invested in mankind's current

way of life.

10. For this reason, it will take numerous generations to turn things around. Therefore, I consider helping young people develop critical thinking skills as being vital for all future generations. This is the topic of Chapter Seventeen in this book.

7.9. Past and Future Universes Are More Likely to Be Physical Instead of Nonphysical

Since the spirit world is nonphysical, and whenever it needs to design and bring into being a new universe to help restore balance the new universe has to have some attributes that are opposite those of the spirit world, the new universe is more likely to be physical than nonphysical. It depends on what caused the imbalance to develop.

The fact that most of the UFOs and UAPs appear to be physical could lend support to this conclusion. They could be carrying visitors from other universes that exist along with our universe, and those other universes are apparently also physical. However, the UFOs and UAPs could also be from other planets in our universe instead.

Some of the UFOs and UAPs that appear as balls of light or patterns of light could be nonphysical. Therefore, this would not rule out that other existing universes could be nonphysical.

Another possibility is that nonphysical UFOs and UAPs could be visitors from other planets in our universe who learned how to do space travel by going through the spirit world.

The presentation in this section of this chapter is to provide some ideas to consider when someday we humans on Earth learn how to do space travel and/or universe travel in a practical manner.

Part Two

Implications of the Spirit World and the Spiritual Model Presented in Reference 1

Chapter Eight: Gradations and Diversity Are Among the Most Important Attributes of the Spirit World

Chapter Nine: Mankind's Gradations and Diversity Constitute a Gift to Mankind

Chapter Ten: Anything Reasonably Formulated Would be Correct but Incomplete

Chapter Eleven: Evolution Spans All Past Universes, Living Things Emerge by Retracing

Chapter Twelve: An Analogy between Evolution and Maturation

Chapter Thirteen: Hang-Ups

Chapter Fourteen: Our Choice of Dominant Basis for How We Carry Out Our Lives

Chapter Fifteen: Combining Spirituality and Science in Our Pursuit of Advancements of Both Kinds

Chapter Sixteen: More Discoveries and Developments the Spiritual Model Could Explain

Chapter Seventeen: The Spirits and Minds of Young People Are Our Greatest Treasure for They Are the Future of Mankind (To Help Young People Develop Critical Thinking Skills)

Chapter Seventeen addresses the importance of developing critical thinking skills in children. This is because the future of mankind is in the hands of today's children. Therefore, we need to do what we can to help them avoid making the mess that members of earlier generations have made.

Chapter Eight
Gradations and Diversity Are Among the Most Important Attributes of the Spirit World

Gradations and diversity are among the most important and valuable natural attributes of the spirit world as revealed by the spiritual model presented in Reference 1. Their importance to us humans on Earth is in regards to how mankind should be carrying out life on Earth as well as perhaps in regards to how life in general should be carried out elsewhere in our universe and in other universes. Both attributes are among many that are brought on by the fact that only one copy of any piece of knowledge is kept by the spirit world.

Because only one copy of any piece of knowledge exists in the spirit world, each piece of knowledge would be shared among countless spiritual entities. It would also result in each smaller spiritual entity to be similarly shared among countless larger spiritual entities. Thus, larger spiritual entities are made of multiple smaller spiritual entities.

Because of these two kinds of sharing, every spiritual entity in the spirit world is directly or indirectly a part of every other spiritual entity. Thus, no spiritual entity would be purely one specific thing. Instead, each spiritual entity is a mixture of bits of numerous other spiritual entities.

What is in the mixture and the nature of the mixture of each spiritual entity would vary depending on the experiences the spiritual entity has gone through. The nature of the mixture could also be adjusted by the spirit world as needed, for example, as it designs a new universe such that the new universe would be able to fulfill a need that developed at the time. Whether or not any adjustment is made, at any point in time the nature of the mixture of any spiritual entity would be located somewhere on a gradation that goes from being purely one specific thing at one extreme to being purely another specific thing at the other extreme.

The location of any such mixture would always be located somewhere between the extremes of any gradation and never at any one of the two extremes of the gradation. This is because in order for a spiritual entity to be located at any extreme of a gradation it has to become purely one specific thing. But it could never become purely one specific thing because it is always a mixture of bits of many things.

A typical spiritual entity being a mixture of bits of other spiritual entities would vary in nature in as many ways as the number of different bits that make up the spiritual entity. Therefore, a typical spiritual entity would be on some location of as many gradations as the number of bits of other spiritual entities it is made of.

The spirit world could adjust the position of any spiritual entity on any gradation by adjusting individually the strength of the signal issued by each piece of knowledge that makes up the spiritual entity and by adjusting individually the transmissibility of each connection of the first kind that also makes up the spiritual entity.

Gradations thus give the spirit world the flexibility to design living things, nonliving things, universes, and anything else to have any configuration, level of mental abilities, acuity of senses, number of major and spiritual senses, etc. as necessary for the living or nonliving thing to have the capability of helping the spirit world restore balance in its state of knowledge.

Because only one copy of any piece of knowledge exists in the spirit world, every spiritual entity is different and unique. Therefore, ultimate diversity pervades the spirit world. This diversity enables the spirit world to design and bring into being a limitless variety of universes and of anything else the spirit world needs to design and bring into being.

Chapter 8

Accordingly, gradations and diversity give the spirit world its creative ability, creative strength, and limitless creative flexibility.

It is in the sense of what is presented so far in this chapter that I consider gradations and diversity to be among the most important attributes of the spirit world as revealed by the spiritual model presented in Reference 1.

What this means is in order for mankind to make the most out of being alive here on Earth, it needs to learn how to apply the various kinds of gradations and diversity in positive and constructive ways regarding how it carries out life on Earth.

In other words, mankind needs to value all variations, including variations in race and gender that exists among humans instead of discriminating against and/or mistreating those who are of certain variations. We are all variations, none of us is purely one way or another, as revealed by the spiritual model presented in Reference 1.

It is clear from life's experiences that variations, including variations in race and gender, have broaden and strengthen the creativity of mankind, just as gradations and diversity have given the spirit world its creative abilities. Creativity is one of the most important attributes mankind can have.

It is also clear that when a wide diversity of individuals is included to participate in a formulation process the process would produce superior results. Logically, it is because a wider range of mental abilities and a greater amount of mental power are being used to execute the formulation process.

A conclusion then is that one of the reasons mankind has made a mess on Earth for thousands of years is because mankind has not wisely applied two of the most importance and valuable human resources that are available to help it carry out life more positively, creatively, and constructively. The resources are gradations and diversity.

Instead, mankind has generally discriminated against, mistreated, and abused such human resources.

It is time mankind gains a much better understanding of the value and importance of these human resources and to develop the wisdom to embrace them and to apply them positively, creatively, and constructively.

The main point here is that mankind needs to be able to form more wisdom, and this can only be done if the knowledge it has about life is more balanced. Right now, its knowledge about life is too heavily weighed on the physical and technical side and too lightly weighed on the spiritual side.

Chapter Nine

Mankind's Gradations and Diversity Constitute a Gift from the Spirit World

9.1. Variations Are Sources of Strength and Abilities for Living Things

The spirit world has developed a process that enables numerous variations to naturally occur with living things. We discussed this in terms of how it works in terms of the spiritual part of life. How it works in terms of the physical part of life is as follows. Keep in mind that what goes on in the spiritual part of life influences how mankind mentally progresses, and what goes on in the physical part of life influences how mankind physically progresses.

At various points in time, a species of living thing would produce numerous off-springs. Because the spirits of the existing living things that produced the off-springs are each a mixture of numerous things as explained in Chapter Eight, numerous new different variations would naturally be produced among the spirits of the off-springs. The spirit of each off-spring would have a unique and different mixture of different things as the mixtures of the spirits of the "parents" are mixed together to "randomly" produce a unique and different mixture for the spirit of each off-spring.

We could see this is the case in our physical world as a typical off-spring would have a mixture of DNAs that are different from the mixture of DNAs of the parents, but the off-spring's DNAs would come from the parents.

This is essentially additional confirmation that the spiritual model presented in Reference 1 is correct at least to significant degree.

This is one of the spirit world's ways of enabling new variations of a species of living thing to evolve from an existing variation when the environment has changed to where the existing variation would no longer be as viable as before. Another variation would then take over to be the most viable in the changed environment. This enables the species to continue to exist and be viable as environments continue to change. Earth's history as recorded in ancient fossils and earthen layers indicates environments will change sometimes quickly but usually very slowly.

One of the variations would be more viable in whatever environment they are in than any of the others. This variation is likely to be more able to reproduce than the others and would thus become the dominant variation for the species for the particular environment they are in at that time.

If their environment were to change, a different variation would become the dominant variation for the species for the changed environment, and the species would thus continue to exist. This would be one of the ways physical evolutionary progressions could take place. It applies to mobile living things such as animals and birds and also non-mobile living things such as plants.

The spiritual model presented in Reference 1 indicates that each time a change in the variation that becomes the dominant one, the species would be advancing more quickly during the change than it would when such a change is not happening. The process would be as follows:

1. If the environment is not changing, the species will be advancing at a normal rate as it continues to go through new experiences and generate new pieces of knowledge that become a part of the spirits of the species. The process would be very slow and any change is not likely to be noticed in the short term but could be noticed in the long term.

2. If the environment is changing significantly and fairly quickly, the species would be going through new experiences at a faster rate in its effort to stay viable. Therefore, new pieces of knowledge are being generated and added to the species' spirit at a faster rate. Thus, the species would be advancing at a faster rate than it would be if the environment were not changing.

3. If the environmental change is slow enough, the species would be able to change quickly enough to remain viable, and it would do so in term of the range of variations it is able to naturally produce. The wider the range the more able it is to remain viable.

4. Each succeeding variation that emerges as the dominant variation following an environmental change is likely to have more physical abilities and stronger mental abilities than the variation that was dominant before had. This is mainly because the spirit of the species has grown from all the new pieces of knowledge that was generated and added to it.

Therefore, the ability to have variations is a source of strength and abilities for living things, according to the spiritual model presented in Reference 1.

5. If the environmental change is too drastic and/or too fast such that the species is unable to adjust quickly enough, the species would go extinct. Such an environmental change could be in the form of a new predator coming into the area, a new living thing competing for a limited food supply, a giant meteor striking the planet and greatly reducing the planet's ability to sustain living things fairly quickly, the axis of rotation of the planet shifting and thus disrupting the environment everywhere almost instantly, etc.

6. On Earth, now that we humans have reached our 21st century, all the species of living things that managed to remain viable are

now quite advanced compared to how they were millions of years ago in terms of their spiritual status, mental ability, and physical makeup. The mechanism by which they have become to be how they are now is likely to be as described in Chapter Eight and in this book.

This then could be yet another confirmation that the spiritual model presented in Reference 1 is correct at least to some significant degree in terms of explaining how variations could naturally happen with living things and how variations plays a major role in spiritual and physical evolutionary progressions.

9.2. The Spirit of Each Living Thing in Any Universe Is a Portion of the Spirit World, and Thus, the Living Thing Could Do Certain Things the Spirit World Could Do

Larger spiritual entities are made up of a large number of smaller spiritual entities. Because only one copy of any existing piece of knowledge is kept by the spirit world, smaller spiritual entities would naturally be a part of a very large number of different larger spiritual entities. Thus, larger spiritual entities could be perceived as being made up of parts of other larger spiritual entities.

A physical situation that is parallel to this spiritual situation exists in our physical world. It is that different living things on Earth are made up of different combinations of DNAs. The same collection of DNAs in different combinations would make up different living things. Therefore, living things could be perceived as being made up of parts of other living things.

This parallel exists since each DNA exists on Earth because a certain smaller spiritual entity enables it to exist. And, each living thing exists on Earth because a certain larger spiritual entity enables it to exist.

Because different living things are made of different combinations of

DNAs, we humans are able to learn how to do genetic engineering and to be able to produce hybrid animals and other hybrid living things. Such hybrids are not found to exist on Earth naturally probably because they would not be the variation that could survive on Earth naturally. Some of such animals and living things appear to be viable at least in a laboratory setting and/or under the care of humans. However, they might not be viable if released into the wild to try to survive on their own.

What this means in my opinion is that we humans, and also everything else in our universe and in other universes, are each a portion of the spirit world as stated earlier in this book. By being a portion of the spirit world each of us humans are capable of doing certain things the spirit world could do, and in this particular situation it is to do genetic engineering. The spirit world essentially does genetic engineering when it designs new living things to reside in a new universe.

We could also say natural cross pollination performed by bees and other insects and also by some birds is a form of genetic engineering. Natural cross pollination caused by wind would then also be a form of genetic engineering. The similarity here is that such natural cross pollinations could be perceived as being performed by certain portions of the spirit world as well, those portions being the bees, insects, birds, and the wind.

In my mind, all this lends support to the notion that the spirit of each living thing that exists in any universe is a portion of the spirit world such that each living thing would naturally be capable of doing certain things the spirit world is able to do. This should not be surprising since the spirit of each living thing is indeed a portion of the spirit world.

9.3. Mixed-Breed Living Things Are Medically Stronger than Thorough Bred Living Things

Each living thing has a different mixture of DNAs; and is thus a different variation of a species. Thus, in this sense each living thing is a slightly different breed, or in the case of us humans, a slightly different race. Off-springs of the same family are going to have mixtures of DNAs that are more alike than are mixtures of DNAs of off-springs of different families.

We humans have learned that a set of parents should not consist of two off-springs from the same family because their off-springs could have various forms of medical weakness. The breeding of thorough bred dogs has shown to possibly produce dogs with various forms of medical weakness.

In spiritual terms, the explanation for how this could happen is as follows:

1. Normally, when two individuals produce a child, the spirit of the child would have a wide range of spiritual strengths from which to pick and choose to form its own combination of spiritual strengths. The available range of spiritual strengths would come from the spiritual strengths of both parents, and it would normally be wide because the two parents would normally come from different families. Thus, each parent would provide a very different range. Therefore, when combined, the combinations of available spiritual strengths for the child's spirit to pick and choose from would be much wider than that of each parent.

2. But, when two off-springs from the same family produce a child, the spirit of the child would have a narrower range of spiritual strengths than normal from which to pick and choose to form its own combination of spiritual strengths. This is because the spiritual strengths of both parents are very similar and their ranges are very similar. Therefore, when combined, the combination of available spiritual strengths for the child's spirit to pick and choose from would be much narrower than normal.

3. A spirit of person is a large spiritual entity, and large spiritual entities are made up of numerous smaller spiritual entities. No two spiritual entities could be identical, because if they were identical they would merge and only one would remain. Therefore, a child's spirit could never be identical with either of the two parent's spirit.

4. When the parents are from two different families, the child's spirit would have no problem being different from the spirit of either parent.

5. But, when the parents are off-springs from the same family, the child's spirit is forced to be similar to the spirits of both parents and at the same time be different from them. Therefore, a way to accomplish this is to make do with a fewer than normal number of spiritual strengths. In other words, the child's spirit would end up being a large spiritual entity that is made up of a fewer number of smaller spiritual entities than normal.

6. There are a large number of smaller spiritual entities that are required to form a human physical body. Therefore, these have to be included in the child's spirit. However, there are certain other smaller spiritual entities that could be left out. What could happen is, for example, a hip joint would be formed well enough, but it might be less durable as normal, and arthritis could develop early.

9.4. Could and Should a Super Human Be Produced Through Genetic Engineering?

We humans wondered if we could produce a super individual or race through genetic engineering. Theoretically it is possible, and perhaps it is better to try doing it in the spirit world first than in our physical world. We would have a better chance at understanding the negative effects of what we are doing. But we would then have to first become very knowledgeable about how things work in the spirit world.

Also, it depends on what super attributes we want to have in a super individual. Chances are slim we could develop an individual who is super in every possible way; i.e., mentally, medically, physically, etc. There is no known agreement as to what a super individual is supposed to be like.

This being the case, in a sense and on rare occasions super humans have developed naturally. They would include (1) an individual who could play the piano expertly without having taken any lessons and having only listened to others who played the piano expertly. (2) Another individual is able to quickly identify the day of the week for any date in the past. This indicates that it is likely possible to produce a super human with a specific super attribute. But, so far it is not known if any human has naturally developed multiple super attributes.

I personally know of several individuals who would get "A" grades in any grade-school and college course they take but who would struggle later on to have a comfortable and happy life. During their childhood they would be perceived as being sort of super-human like, at least academically, but in their adulthood their super-human abilities ended up being taken advantage of by others to enrich themselves.

Another observation is that we humans with our current state of human nature are prone to doing bad applications of our technical advancements. This means in my opinion that if we were to find a way to develop super humans in terms of any attribute, we are likely to find a way to do bad applications of our ability to do this.

Therefore, I think we human should not try to find a way to develop super humans until after we have learned enough about the spiritual part of our lives such that we are able to form enough wisdom to stop doing bad applications of our technical advancements.

9.5. Mankind's Gradations and Diversity Constitute a Gift to Mankind from the Spirit World

Explained in Reference 1 is that the more advance is a species of living thing the faster that species would become more advanced. The human species on Earth is an example of this. The more advanced we humans become, the more complex would be the experiences we could construct for ourselves to go through and thus the more complex would be the kinds of new pieces of knowledge we would generate, and the faster we would generate them.

This is how the human species got to be so much further ahead of every other species of living thing on Earth in terms of our evolutionary progressions.

In addition, with our high level of imagination and creativity we would also become increasingly technically advanced such that we could invent all kinds of machinery and tools to leverage our abilities. Therefore, we could build amazing things that we would not be able to build with our bare hands. We could also find ways to enable ourselves to be safe and viable in just about anywhere on Earth.

This is a major difference between the human species and the species of other living things. We humans have been using our superior mental abilities and creativity to enhance our chances of survival no matter the environment in which we are in. We invented clothing, dwellings, and vehicles to keep the climate that is right next to our bodies essentially the same no matter the natural environment that surrounds us. Therefore, we have essentially eliminated most of the effects the environment has on our chances for survival.

However, different natural environments would still require different physical and mental abilities, different skills, different kinds of dexterities, and different kinds of immunity strengths, etc. for our survival. This is because different natural environments present different kinds of experiences we must go through to survival. Therefore, how our spirits would grow and develop would be different in different natural environments.

The result is the human species developed numerous different races, each with a different but equally valid and equally valuable set of abilities and strengths. Most important of all, no contemporary human race has ever gone extinct, and this is a major difference between the human species and all the other species of living things on Earth. All human variations are viable and are present whereas for most other species of living things only some variations remain viable and the rest have gone extinct. Exceptions would include domesticated animals.

While all human races are equally viable and equally strong, the wonderful thing about the human species is that each race is strong in a different way and all races managed to be viable on Earth. This means the gradations and diversity that is embodied in the large spiritual entity that enables the entire species of humans to exist on Earth managed to be translated in a manner that enabled all the variations of the human species that have ever existed are in existence in our physical world today. This is true mostly for the human species, whereas most other species of living things have only a few of their variations that ever existed are in existence today.

What this means is that our entire collection of human variations (races) is viable and exists on Earth today to work together to broaden and enhance mankind's creative power, mental ability, mental strength, and a variety of various other abilities. It is in this sense and the fact that the gradations and diversity that exist in the spirit of the entire human species is what enables all variations of the human species to be viable and here today that I say the following:

Mankind's Gradations and Diversity Constitute a Gift to Mankind from the Spirit World.

This is supported by the fact that it is quite clear from mankind's experiences on Earth that the following is true:

When a wide gradations and diversity of individuals are included to participate in a formulation process the process would produce superior results.

While having this gift is wonderful for us humans, however it is hardly taken advantage of by us humans on Earth. Instead we humans tend to discriminate against and to abuse what this gift is able to bring about. Such behavior is extremely unfortunate because instead of making progress toward fulfilling mankind's primary purpose, mankind has made a mess on Earth. It is because mankind has not gained the wisdom to realize what it has is this wonderful gift from the spirit world and that it should be used it in positive, creative, and constructive ways.

It is also mankind's development of large egos on top of a lack of wisdom that resulted in such poor overall behavior. Therefore, in addition to doing what is necessary to increase mankind's wisdom, mankind need to also do what is necessary to significantly reduce its ego. I think mankind could make significant progress in both areas by seriously pursuing spiritual advancements.

In the beginning of this section, I explain how the human species advanced so much further than any other species of living things on Earth. However, our advancements have been mostly technical while our spiritual advancements have lagged far behind. It might very well be possible that some of the other species of living things on Earth have advanced spiritually much further than the human species has. After all, we do not see any of them mistreating members of their own species as badly as we humans do, or that they make wars with members of their own species. When we look at it that way, the human species does not appear to be very advanced.

9.6. Gradations and Diversity Could Always Enable Mankind to Come Up With a Superior Version of Just about Anything

Since gradations and diversity are a natural attribute of the spirit world, they are bound to be a natural attribute of every living thing the spirit world creates. This notion is supported by the gradations and diversity of the DNAs making up living things. In the case of us humans, this notion is made especially obvious by the large number of human races that exists on Earth. As discussed earlier, except for domesticated animals, most other species of living things on Earth would typically have only a few of their strongest variations remaining to exist while the rest have gone extinct.

Since each human race developed its own way of being viable while residing in an environment that is different and unique for the region in which the race resides, each human race would have a different and unique outlook on life. This means if a specific human race were to formulate a spiritual model, it would be different and unique from a spiritual model formulated by any other specific human race.

In general, this appears to be the case as indicated by the various religions, traditions, customs, and ways of life that have developed by the various human races.

According to the spiritual model in Reference 1, every model reasonably formulated would be correct but incomplete. What this means is any spiritual model reasonably formulated by a specific race of humans would be correct but incomplete such that if we were to combine all such models together from all specific races of humans we would come up with a combined model that is correct but more complete than any one of such model would be.

This would hold true for any kind of model or anything that the various human races would reasonably formulate. The combined version of anything would be more complete than any one version would be. This is one of the ways that mankind's gradations and diversity would directly constitute a gift from the spirit world. It is because the human species would always be able to come up with a correct but more complete version, and thus a more superior version, of just about anything.

9.7. Taking Science into Consideration to Formulate More Complete Spiritual Models

As far as I can tell, the spiritual model presented in Reference 1 is the first to take into consideration engineering reasoning in its formulation. The model turns out to be capable of providing logical, reasonable, and explicit spiritual explanations for essentially all common everyday observations and experiences we could think of.

The model therefore points out the value and importance of including science related methods in the formulation of spiritual models. It enables the resulting spiritual model to be more complete in terms of capturing more of the attributes about life. Thus, the spiritual model would be more effective in inspiring and motivating mankind to improve itself because the will to do so would originate internally within each individual. This is bound to work better than externally imposing laws and rules upon each individual in an attempt to confine mankind into behaving better.

The feeling of confinement of any kind is bound to motivate mankind to find ways to break out of the confinement. In this case this could be in the form of breaking the laws and rules. An example of this in years 2016 through 2020 is how it is that the president of the United States of America is managing to keep the loyalty of his base supporters in spite of how he is breaking all kinds of laws and rules. It is possible that his base supporters are getting a lot of satisfaction from his behavior and are therefore more than willing to accept and justify how he is behaving.

This could also explain how cult leaders could be successful cult leaders, and how followers could become staunch cult followers. If this is true, then it points out a need for a more effective way to get mankind to behave better than simply externally imposing laws and rules. We need to develop ways that would inspire and motivate individual from within themselves to behave better.

A way to do this is to enable mankind to gain a more complete knowledge about life. Whatever would be a workable way would likely take many generations to accomplish in order to have any degree of permanency. Any quick way is likely to work only temporarily and only under controlled conditions. History has shown this to be the case.

The applications of mankind's technical advancements are spreading across all human races. This means an increasing number of human races are getting used to applying technology while carrying out their lives. This means an increasing number of human races are in a position to formulate spiritual models that could take into consideration science related methods.

Therefore, the potential exists to have available a diverse collection of spiritual models that are capable of providing a more complete and thus a more logical, reasonable, and explicit description of common everyday life. Each such model would be doing this in a different and unique way because each human race would have a different and unique perspective about life. Thus, the combination of all such models would come up with a model that is more complete and more effective than any one such model could be. As discussed earlier, this is why gradations and diversity constitute a gift to mankind form the spirit world.

One of my hopes is that the spiritual model presented in Reference 1 would encourage others to formulate new spiritual models in which the formulation process takes science related methods into consideration.

Because the human species is made up of a diversity of different races, mankind has the potential to come up with a very complete combined spiritual model that is made up of numerous spiritual models each of which takes into consideration science in a different and unique way.

It would be beneficial to mankind to encourage such formulations of new spiritual models. A diversity of such spiritual models would enable mankind to understand life more completely. Such spiritual models would also be more appealing and more relevant in today's world in which the world population is becoming more educated and where technology is playing an increasingly role in everyone's life. Accordingly, we need to have more spiritual models in which science is taken into consideration in their formulations.

Because I am an engineer in my profession, the formulation of the spiritual model in Reference 1 takes into consideration engineering logic. It would be great if individuals in other technical professions, especially professions in various science-related fields, would formulate spiritual models while taking into consideration science-related concepts in their formulations.

Chapter Ten
Anything Reasonably Formulated Would Be Correct but Incomplete

10.1. Incompleteness Is a Natural Attribute of Everything

Every existing piece of knowledge is correct. Every spiritual entity is correct since it is made of pieces of knowledge all connected together. The spirit world, being the largest possible spiritual entity at any point in time, is correct and would always be correct as it continues to grow as new pieces of knowledge are generated and added to it.

However, as stated earlier in this book, the spirit world would always also be incomplete, because it could never possess every piece of knowledge that is possible to generate. No matter how many pieces of knowledge have already been generated, additional new pieces of knowledge could always be to generate, because new experiences could always be constructed.

This is analogous to how it is always possible to form a new word or add a new meaning to an existing word no matter how many words already exist. We can see how this is the case whenever we achieve a highly significant new technical advancement. We would form new words and/or add new meanings to existing words to describe something associated with the advancement. For example, the term "artificial intelligence" did not exist or did not explicitly mean anything until technical advancements made artificial intelligence possible. The same goes for the following terms and words: social media, vaping, quantum computing, selfie, genetic engineering, etc.

New experiences are made possible whenever a new piece of knowledge is generated and added to the spirit world. The number of spiritual entities in the spirit world would increase approximately two times an exponential increase, as explained earlier. The spirit world is thus changed from how it was before the piece of knowledge was added, and new experiences that are different from any that were previously created would now be possible. The new experiences in turn would enable more new pieces of knowledge to be generated and added to the spirit world. It is a vicious circle.

Consequently, it is impossible to ever run out of new pieces of knowledge possible to generate. And, therefore, the spirit world would always be correct but also always incomplete. This means anything the spirit world creates would always be correct but incomplete as well, including anything that could be translated to exist or be expressed in our physical world.

Therefore, nothing that exists in our physical world could ever be perfect. This explains why we have never seen or heard anything that is perfect, and why what might be considered perfect by one person would not be considered perfect by another person.

We need to also point out that while any specific thing the spirit world creates is correct, the application of that specific thing could be incorrect. Under such circumstances, for expediency, we would often say that a specific thing is incorrect or wrong. Although, it is more accurate to say it is the application of that specific thing that is incorrect or wrong but not the specific thing itself. More discussions regarding such circumstances are presented later in this chapter.

The indication that the spirit world is always correct but incomplete was made evident by the explanation provided by the spiritual model presented in Reference 1 on how this is the case.

It is also evident that a spiritual model that includes science associated considerations in its formulation would be able to provide a

significantly more complete description of life than could spiritual models that do not include such considerations in their formulations, because science is a part of life as well as spirituality is a part of life. Conversely, science should take into consideration spirituality in its model formulations as well when possible and reasonable to do so.

10.2. The Driving Force Behind Why Mankind Keeps Doing Research

As pointed out, it is more accurate to say something that is reasonably formulated to be correct but incomplete than to say it is wrong. This becomes clearer if we are able to observe what is going on in the spirit world while we are reasonably formulating something. According to the spiritual model in Reference 1, the follow would be happening in the spirit world when we are formulating something in a reasonable manner:

1. We would use our spiritual senses to first carefully select all the spiritual entities that are appropriate for formulating the certain something we plan to formulate.

2. However, because the spirit world is always correct and incomplete, the collection of spiritual entities that are available to do a formulation would be incomplete and will always be in complete. Therefore, the formulation we are able to do would be incomplete at this point in time, and it would be incomplete also at any other point in time.

3. This is because at any point in time there will always be additional spiritual entities that are possible to be created that would be applicable for doing the formulation at hand that are not yet possible to create because the pieces of knowledge to create them are not yet generated. Therefore, at any point in time the spirit world would not have all the pieces of knowledge necessary to create all the spiritual entities that we like to have

at that point in time.

4. At the next point in time that is of interest to us, additional spiritual entities would have been created because some additional new pieces of knowledge would have been generated and added to the spirit world. A different incompleteness would exist, but it would be of a nature that is similar to the nature of the incompleteness that was there before. Therefore, there is actually no end to the number of spiritual entities that we would like to have available, because with each new one created, we would wish additional new ones would be available.

5. Consequently, such a situation will always exist in the spirit world and that is why the spirit world is always correct but in complete, and that is why anything we humans reasonably formulate would always be correct but incomplete as well.

6. This perpetual state of seemingly continuing improvement in in completeness is the driving force behind why mankind keeps doing research. It is sort of like an addiction, because while the incompleteness seems to be continuously improving, it is in reality also growing as the spirit world is growing even as it seems to be improving. Therefore, the incompleteness simply never ends. However, researchers are enjoying the discoveries they keep on making each step along the way.

7. Thus, the process is self-perpetuating because the more we know the more we want to know. This could explain why when re searchers discover something, the discovery would lead to more things to be discovered. Thus, our pursuit of technical advancements keeps on going. The driving force behind all such actions is because the spirit world is always growing but staying correct but incomplete. This means such an action is likely to also happen when we start pursing spiritual advancements in a manner similar to how we pursue technical advancements.

8. We might then ask: wouldn't mankind eventually make all the discoveries that are available to make such that the process would eventually reach an end? The answer is that an end doesn't seem to exist. I think one reason is because as mankind continues to make discoveries, it is coming increasingly close to the interface between our physical world and the spirit world. Once we go past the interface, we would be researching into the spirit world.

9. A possible indication of this is how one researcher who is attempting to identify the smallest possible physical thing once said that the smallest possible physical thing seems to take on the qualities of a thought. According to the spiritual model presented in Reference 1, such a possibility appears very likely to be case.

10. Once mankind has gone into the spirit world and start to research the spirit world, the discoveries that could be made would be never ending. One reason is new discoveries would keep being created as new pieces of knowledge are generated and added to the spirit world.

11. An interesting possibility about such a continuing research effort is that it could eventually lead to mankind finally realizing life on Earth has a spiritual part and a physical part. When that happens, the technical models that mankind has been dealing with will be broadened to include spiritual considerations. This would be arriving at combining science and spiritual consideration from the side of technical model formulations instead of the side of spiritual model formulations.

12. When that starts to happen, it would be great to then reach for combining science and spirituality from both sides at once such that each effort could enhance the other. The result is bound to be superior to doing the reaching only from one side or the other.

13. If what is speculated could be real, then it means mankind will eventually mentally evolve to where it will have a more complete knowledge about life. Mankind would arrive at this by its continuing pursuit of research. However, the process is likely to take considerably longer to get there than if we were to do what we could do now, and that would be to include science considerations in our formulation of spiritual models.

10.3. Better to Say "It Is Correct but Incomplete" Than to Say "It Is Wrong"

We tend to say something is wrong when it doesn't match what we consider to be right. However, if that something is reasonably formulated it is bound to be at least somewhat correct even though also incomplete. But we tend to see any incompleteness as being wrong. This could bruise egos even if only slightly, and then some kind of battle could ensue even if only mildly. In a severe situation a severe battle could ensue.

On the other hand, if we say something is correct but incomplete instead of saying it is wrong, it is not likely to cause a battle. Instead, the parties involved might want to get together to see if they could come up with something correct and more complete. I.e., it tends to promote peaceful collaboration instead of battles.

Besides, it is more accurate to say something reasonably formulate is correct but incomplete than to say it is wrong.

It could also motivate people to see what might be important that was left out of their formulation that could now be included. By having such a motivation initiated, the something could become more complete if reformulated. This would then make communications and interactions with others go more clearly, smoothly, effectively, and thus more constructively.

10.4. Saying Something Is Either Right or Wrong Is More Expedient for Some Situations, Even if Less Accurate

In certain technical fields it is more expedient and therefore perhaps more appropriate to say something is "right" if the effects of its incompleteness are low enough to be acceptable, and to say it is "wrong" if the effects of its incompleteness are too high to be acceptable. Examples of technical fields in which this applies are structural design, medicine, food processing, food preparation, children's toys, etc. These fields produce things that could be a danger to people if the things are not formulated correctly. Thus, it would be "right" to allow people to buy and use a product if it is safe, and it would be "wrong" if it is unsafe.

Theoretically it is more accurate to say the design of a product is correct but incomplete if the effects of the incompleteness are low enough to be acceptable or is incorrect if the effects are too high to be acceptable. But where safety is a concern no one would want to buy and use a product in which its design process is theoretically incomplete even if the effects of incompleteness are acceptably low. Therefore, it is appropriate and more expedient to say the design process is either right or wrong, and also that it is either right or wrong to allow people to use the product.

As an example, here is what structural engineers would do to design a structure:

1. They would apply linear theory of elasticity to characterize the structural properties of the building materials to be used.

2. They would know the actual behavior of all materials is nonlinearly, and that the error in applying a linear approximation would be acceptably small if applied within certain limits. The limits would be set such that the error would stay below an acceptable level.

3. They would also know every material will fail at some level of stress and/or strain. The maximum stress and/or strain allowed

would be set way below their failure levels.

4. These limits would be met by incorporating appropriate safety factors and/or safety margins in their design analyses.

5. The structural design would thus be correct but incomplete. The incompleteness is because the true effects of nonlinear behavior are not determined by their design analyses. The nonlinearities are left out of their design analyses because to include them would make the analyses so complicated that the analytical results would be unreliable.

The mathematics involved with nonlinearities of any kind is extremely complicated, and the process of getting a solution for the messy equations involved would require making numerous approximations in order to make them tractable. At the end the engineers would not know for sure what the analytical results mean. Thus, the engineers would use linear theory instead and they would make sure the effects of the incompleteness would be low enough to be acceptable by applying appropriate safety factors and safety margins.

6. In this case, it would be more expedient and more appropriate to say the design analyses are "right" and that it is also "right" to allow the structure to be used by people because it is safe to do so.

And conversely, it would be more expedient and more appropriate to say the design analyses are "wrong" if the design analyses were done incorrectly and that it would also be "wrong" to allow people to use the structure if it is unsafe to use it.

After all, it would be too complicated and confusing to say the design analyses and the structure are correct but incomplete such that the effects of the incompleteness are low enough to be acceptable. Any potential user of the structure would not understand why something that was not completely designed and not completely constructed could be safe to use even though describing the design analyses and the structure in this manner would be more accurate.

10.5. Would Things Unreasonably Formulated Be Incorrect?

We might ask, why do we say anything "reasonably" formulated would be correct but incomplete, which then implies things "unreasonably" formulated would be incorrect. But how could that be, when all spiritual entities are correct? After all, things could exist in our physical world because spiritual entities enable them to exist. So, if all spiritual entities are correct, then how could some of the things they enable to exist in our physical world be incorrect just because they are unreasonably formulated?

The answer is, when we say something is "reasonably" or "unreasonably" formulated we are talking about formulations made by highly intelligent living things on Earth, such as us humans. We are not talking about formulations made by the spirit world. For example, the process we humans typically go through when we formulate something is as follows:

1. We would first imagine the various things we could assemble to formulate the something we have in mind. More specifically, we would use our spiritual sense to find the spiritual entities which are the spiritual forms of these various things.

2. We could imagine these things only if their spiritual forms already exist in the spirit world, and they would all be correct in the spirit world. There are likely to be things that we could not imagine, because they do not exist, that could be important to our effort to formulate the something we are attempting to formulate. And, we would not know that their nonexistence

could be important to our success with our formulation process.

3. In addition, we would be able to assemble the things to form that something only if the spiritual form of that assembled something already exists in the spirit world. In other words, we do not actually create that something by formulating it and mentally assembling it. Instead, we would find the spiritual entity in the spirit world that is the spiritual form of that assembled something such that we would be able to assemble that something.

If the spiritual form of that something does not exist in the spirit world or if we are unable to find it, then we would not be able to assemble the something that we want to assemble. Or what we have would turn out to not be able to be assembled into being that something.

4. Next, if we were able to find the spiritual entity that is the spiritual form of the something that we formulated we would then translate that spiritual entity into a form that could exist in our physical world. That is when we would consider ourselves as having been able to formulate that something.

5. Then the next question is, does the translated form make sense for what it was intended to do. If it makes sense, then it was reasonably formulated. Otherwise it would be unreasonably formulated.

6. For example, what is reasonable and correct in one culture could be unreasonable and incorrect in another. And what is reasonable and correct for one purpose could be unreasonable and incorrect for another.

7. Therefore, whether a formulation is reasonable or unreasonable is in the hands of the individual doing the formulating. This means that individual needs to be knowledgeable about a wide

range of things and also be able to think critically in order to be able to be sure the formulation would be reasonably done.

10.6. Formulations Could Be Correct but Inappropriately Incomplete

Have you ever wondered why good individuals would sometimes purposely break the law or unintentionally get into trouble? This happens when they formulated decisions that are correct but inappropriately incomplete.

We tend to perceive such incidences as being wrong instead of correct but incomplete. By saying an incidence is wrong we are likely to bruise that individual's ego, and this could create other problems such as the individual blowing off steam at something. After all, that individual is likely to feel bad already and angry at himself or herself for having caused the incidence.

On the other hand, saying the incident is correct but incomplete would be more constructive and it would tend to leave the individual's ego intact. It could motivate the individual to think about what he or she should have included in the formulation of his or her decision.

For example, suppose an individual decides to drive way above the speed limit. The decision would be reasonably formulated if the only things necessary to consider are that everyone else is speeding and no highway patrol officer is around. But, two important things were left out in the formulation, and that made the formulation inappropriately incomplete and thus the decision incorrect. They are (1) the danger speeding poses to others and to himself or herself, and (2) his or her speeding could encourage others to also speed and thereby further increasing the danger to everyone around.

In this case, the decision to speed could be either considered 'wrong" or be considered "correct but incomplete." We could consider it to be wrong because the driver is violating the speed limit law. But then this means the decision that practically every other driver on the same road made to violate the speed limit law is also wrong. Then we could go further to say "not enforcing the speed limit law is also wrong." Therefore, the whole situation that exists at that point in time on that road is wrong.

However, looking at it this way is unlikely to motivate any of the drivers to rethink what they are doing and to make a better decision. In fact, they are more likely to justify their decision to break the speed limit law because every other driver is doing it.

Alternatively, I think it is better to say the decision to speed is correct but incomplete. It is correct because every other driver is speeding and no highway patrol is in sight. But it is also incomplete because the drivers did not take into consideration that speeding is breaking the law. I think this would have a better chance of motivating the drivers think about what is lacking in their decision-making process. Therefore, they might rethink what they are doing and thus make a better decision.

I think if we seriously pursue spiritual advancements such that our knowledge about life would become more complete, we would become less likely to formulate decisions that are correct but inappropriately incomplete. For example, in case of the situation discussed, by becoming more knowledgeable about the spiritual part of our lives we are likely to be more concerned about how speeding would endanger everyone around and that our speeding could encourage others to speed as well, and that would make the danger worse.

10.7. Dealing with Gradations as Gradations vs. Dealing with Absolutes

Because gradations are simply a nature of how things are in the spirit world, the concept that things are "correct but incomplete" fits right in, because it is in essence a description of gradations. This is because the incompleteness of things ranges from minimum to maximum, but never totally complete or totally incomplete. In other words, it is on a gradation that goes from minimum to maximum.

For individuals who prefer to deal with absolutes, the uncertainties of this concept could create some confusion and problems even though it is an accurate description of a natural attribute of the spirit world.

In my mind the ability to deal with gradations is a part of being able to think critically. This is because, the ability to think critically when dealing with gradations is important because gradations are a part of just about everything we have to deal with in life.

However, dealing with gradations takes a lot more effort, time, and patience than dealing with absolutes. So, for expediency we tend to handle a lot of gradations as if they are absolutes, and then we might wonder why what we are doing doesn't always work well. It is because what we need realize is that what we are dealing with are gradations, not absolutes, and that the perception that something being an absolute is a man-made thing. We are more likely to break laws that we perceive as absolutes than we are if we were to perceive what is right or wrong as being on a gradation between absolutely right and absolutely wrong.

For example, we set laws and rules as a way to deal with gradations. And then we tend to break laws and bend rules when think we can get away with it. People tend do that more often while driving a car.

Laws and rules are examples of how someone other than us is doing the thinking for us. In a lot of situations laws and rules are necessary in order to have things work at all. In my opinion this is because we humans are missing the spiritual part of the knowledge that we need in order to fully understand life. I suspect if we humans were to have that part of the knowledge about life, we would be able to handle gradations quite well, and a lot fewer laws and rules would be needed.

We humans could gain such knowledge by pursuing spiritual advancements. But it is going to take quite some time to gain enough of such knowledge to enable us to be able to deal with gradations well. However, meanwhile we could help young people develop critical thinking skills so that they would be more able to deal with gradations in addition to numerous other problems they will face in life. But, in my mind, the ability to deal with gradations well is one of the most important abilities people in general need to have.

Esther Wojcicki, the author of the book "How to Raise Successful People", Reference 7, has a good method for teaching children to deal with gradations. She does not say gradations are what her method deals with, the word "gradation" is what I use, but her method deals with gradations in my opinion.

She gave the example of how to stop a child's tendency to use their cell phones during class time. Usually, adults set the rules, and the children are supposed to follow them. But in such situations the children usually try to find ways to break the rules if they can get away with it.

In this case Esther sets the goal of zero use of cell phones during class time, and then let the children decide how to go about meeting the goal. Her method worked much better than usual, because the children had a role in deciding what to do. Therefore, the inspiration and motivation to follow the rules originated from within themselves as opposed to having rules imposed upon them from outside of themselves.

What Esther did was to acknowledge gradations exist in that every child is different and every situation is different. Thus, using absolutes as a way to handle gradations is not going to work well in general. But setting the goal and then letter the children decide how to meet it would be more effective when dealing with gradations in general because that approach acknowledges that gradations exist.

She uses different terminology to describe how her method works, and it is different from my terminology. This just goes to show there are multiple ways to describe something. Describing her method from multiple points of view provides a more complete understanding of her method than describing it from only one point of view. This is like describing the spirit world with numerous different spiritual models would provide a more complete understanding of the spirit world than would be provided by just one spiritual model.

10.8. The Spiritual Model Presented in Reference 1 Is Correct but Incomplete

It follows from the presentation so far that the spiritual model presented in Reference 1 would be correct but incomplete just as every other spiritual model would be. Therefore, while it enables us to start pursuing spiritual advancements in a manner similar to how we pursue technical advancements, it is only a beginning. As we get more into achieving spiritual advancements we are bound to naturally come up with additional spiritual models, ones that are more complex and more complete than the one presented in Reference 1.

As an example, what is presented in this book already includes a lot of new discoveries and developments that were made by applying the spiritual model in Reference1 after Reference 1 was published. Therefore, in a sense, what is presented in this book is a more advanced and more complete version of the spiritual model presented in Reference 1.

Therefore, such progressions are bound to happen, and have happened and will continue to happen with our pursuit of spiritual advancements just as how they continue to happen with our pursuit of technical advancements. More specifically, each such progressive step enabled us to discover a new capability in our physical world which we would then serve as a platform to formulate a new and more advanced and more complete model to describe how the new capability works. The new model would than serves as a next platform that enables us to launch off to find the next new capability in our physical world. As mentioned earlier, our models are essentially documentations of our discoveries and our technical advancements. This would similarly be the case with our spiritual advancements.

But models are only approximations of reality. Every model has its limitations in that they are valid and reasonably accurate only within a certain range of various parameters. This is because every reasonably formulated model is correct but incomplete and it is the level of their incompleteness that determines the level of their limitations. It is important to always keep this in mind regarding any model we formulate in order to be able to do a reasonable job of formulating the next more advanced and more complex model.

If we were able to formulate "the model of everything", it would have no limitations. However, the spiritual model presented in Reference 1 indicates that it is not possible for us humans to formulate the model of everything in spite of the fact that Einstein tried to formulate it. The next section explains why.

10.9. The Model of Everything Is Not Possible to Be Formulated by Humans

It is more correct to say we have yet to formulate "a" model of everything rather than "the" model of everything. Einstein tried to formulate "the" model of everything but wasn't able to do so. Here are some reasons why:

1. A model of everything is bound to be unimaginably complicated such that the time and effort to formulate it would be so enormous that it is beyond what we humans could do, except possibly Einstein.

2. Its complexity would require a lot of complex thinking to work with it. Einstein is likely capable of such complex thinking, but he might have run out of time.

3. According to the spiritual model in Reference 1, the spiritual form of "the" model of everything would not possibly exist in the spirit world. However, the spiritual form of "a" model of

everything would exist in the spirit world, but each spiritual form could exist only for an instant.

4. The spirit world is constantly growing and changing. This means a model of everything would exist for only an instant and then a new model of everything would immediately replace it every time the spirit world changes; i.e., every time a new piece of knowledge is generated and added to the spirit world. Therefore, there is no "the" model of everything, but rather there is a stream of an ever changing "a" model of everything with each existing for only an instant.

5. We could also say a model of everything is constantly growing and becoming more complex such that any human attempt to formulate it would be extremely difficult if not hopeless. And, if a human were to succeed in formulating an instantaneous version of it, it would immediately become obsolete.

6. According to the spiritual model in Reference 1, even if a human's ability to formulate a model of everything were able to keep up with its extremely fast rate of change, the versions the human is able to formulate would only be a portion of a more complete model of everything. The formulated versions would only be the portion that pertains to our particular universe. Such a model would not include the rest of the spirit world or any of the other universes that might be in existence along with our particular universe. A human would not be able to figure out what the rest of a more complete version of a model of everything would be like in order to formulate it.

10.10. A Crazy Quilt Made of Models and Has Lots of Holes

It would be great if we have "the" model of everything that is also simple to use such that we need to deal with just one model to handle everything in our lives. But as indicated in the preceding section "the" model of everything is not going to be available to us on Earth or to any other living thing anywhere else, especially not one that is simple to use. Therefore, as explained in Reference 1, we are using a "crazy quilt" of models made of all the models we humans use to carry out our lives, and that crazy quilt has lots of holes that are not addressed by any model. If we were to do something that gets us onto one of those holes, that is when we would really get into trouble because there would not be a model to guild us on what to do.

Each model making up the crazy quilt is correct but incomplete. Each model has its individual region of applicability, and its accuracy falls off as we near the limits of its region of applicability. If we need to go further than what a specific model covers, we need to jump off this model onto an adjacent model whose region of applicability begins where the region of applicability of the first model ends. We go from one model to the next repeatedly as we go about doing all the tasks that we need to do each day of our lives.

Over the past many thousands of years, we have formulated a lot of models that are generally fairly easy to use and that could cover every region of life that are important to us. If we connect all these models together, they would form a crazy quilt of models. This crazy quilt would essentially be a very rough approximation of a model of everything. Therefore, instead of dealing with a single hopelessly complicated model of everything we are dealing instead with a countless number of fairly-easy-to-use smaller models and accepting the limitations that come with them.

This crazy quilt of models has holes not covered by any of the models that we humans use in our lives. The holes are where two or more models are supposed to meet but they don't quite meet and thus a hole is formed. Such holes are hazardous because if we try to apply any model beyond its region of applicability, we could end up on one of these holes and thus we are likely to do the wrong thing and run into trouble because we have no model to guide us.

An example of a hole in the crazy quilt would be a disease for which we have not been able to find a cure. Another example is our inability to stop global warming from progressing further and our inability to know what we should do now to protect ourselves from the emerging negative effects of global warming. A rather serious example is our inability to know how to stop criminal acts that are taking place through bad applications of social media.

We might ask, could a model of everything avoid bad application of our technical advancements and could it enable us to find a cure for diseases, etc.? The answer is yes for both questions, provided we humans are able to handle its extreme complexity. It would take a living thing with a metal ability that is way greater than that of us humans. Such a living thing would likely have enormous wisdom and would not need a model of everything to know what to do and what not to do.

Regarding social media, now that we are realizing all the bad applications that are happening with it, I am wondering if its negative side outweighs its positive side. It allowed criminal behavior to become more anonymous such that they are harder to detect and harder to stop. It is as if social media is a nonphysical world that exists in a manner that is parallel with our physical world. Criminals are able to do their bad deeds in this parallel world, and we in our physical world are having a great deal of difficulty reaching into this parallel world to do much to stop such criminal acts.

10.11. Statements Could Be Interpreted in Different Ways Because They Are All Incomplete Even if Correct

Notice how when a speaker makes a statement, different listeners would interpret what was said differently even when the statement is reasonably formulated. One of the reasons this can happen is because every statement reasonably formulated is correct but incomplete. The incompleteness is what enables different interpretations to be made.

Different interpretations can contribute to making a mess in our physical world especially when they are concerning national and international issues. For example, there appears to not be a consensus as to exactly what started World War One. Could this be because various statements that were made at the time were interpreted differently by the people on different sides of the issues of that time?

Incompleteness will always be present in any statement. However, the cause of different interpretations happening is not strictly due to the incompleteness. It is also due to the different hang-ups that different listeners have. Hang-ups, whether they are minor or major would influence how a listener interprets what he or she hears. The influence could be minor or major, and it is often subconscious.

Different kinds of hang-up would be developed during different times. For example, the kinds of hang-up individuals are likely to develop when a civilization is young and just getting started would be different from those likely to be developed when a civilization is old and values have changed. Changes in hang-ups could be another way to explain why civilizations would eventually become nonviable and therefore eventually end.

Incompleteness in statements could be minimized but never avoided. But hang-ups could be resolved. This means if we humans were to resolve hang-ups as much as possible, the differences in the interpretations of statements would be reduced. This would make communications better and more effective.

Hang-ups could be resolved easier and quicker if we humans had a more complete knowledge about life. We are currently lacking in the knowledge about the spiritual part of our lives. Again, we need to seriously pursue spiritual advancements in order to gain this knowledge.

Chapter Eleven

Evolution Spans All Past Universes; Living Things Emerge by Retracing

11.1. The Spirit World Periodically Brings New Universes into Being and Uses Retracing to Have Living Things Emerge in a New Universe

As discussed earlier, anything that lives and grows for a very long time is likely to need periodically adjustments to restore balance and maintain viability. This is especially true for the spirit world since it lives and grows forever. One of the ways the spirit world restores balance in its state of knowledge is to design and bring into being new universes that could generate the kinds of new pieces of knowledge that would help restore balance.

Various attributes of our particular physical universe indicate the spirit world has done this numerous times. For example:

1. **The level of sophistication, creativity, and cleverness in the design of our universe and of the living things that reside in it:**

    This is evidence that the spirit world must have had a lot of experience designing universes and living things. For example:

    a. The basic physical building materials making up just about everything in our universe consist of only three kinds of things; electrons, protons, and neutrons.

    We could go deeper and say subatomic particles are the basic physical building materials. But researchers are still studying how many varieties there are, how small they could be, and all the things they are capable of doing. Therefore, for now we will simply say electrons, protons, and neutrons are the basic building materials of our particular physical universe and of all the living and nonliving things residing in it.

b. It takes a lot of experience, knowledge, cleverness, and creativity to design just three kinds of basic building material that seem so simple and yet are so capable of doing an enormous number of amazing things such as the following:

- Form the enormous variety of living and nonliving physical things that make up our universe ranging from tiny grains of sand to enormous galaxies with giant black holes at their centers.

- Produce the numerous kinds of energies that are a part of our universe.

- Produce the numerous kinds of biological processes that enable living things to be alive and growing.

- Enable all the natural phenomena, actions, and reactions to take place throughout our universe ranging from simple chemical reactions to the massive fusion actions that take place in stars and in our Sun.

c. In addition to the basic three kinds of building materials, our universe includes some non-tangible materials such as dark matter and dark energy. What these materials are and what they do are yet to be determined.

I speculate they might be what the outer space of our universe is made of and that they also produce the gravitational force in our universe. My speculation is based on the following:

- Our particular universe is a physical one. This means it would consist totally of physical things and not a combination of physical and nonphysical things.

- The outer space of our universe has to be a part of our universe. It is unlikely to be the nothingness that was there before our universe was brought into being as I think some of us tend to think.

- Therefore, if the outer space of our universe was brought into being as part of our universe, it has to be a physical thing and not a nonphysical thing.

- Observed phenomena such as, (1) the gravity of a planet has been observed to warp space, and (2) gravitation waves have been observed in space, would indicate outer space is a physical thing and therefore must be made of something physical. I speculate that outer space might be made of dark matter and dark energy

- Dark matter and dark energy might also be what produce gravitational force that holds our universe together.

- Very old, moderately old, and very young galaxies are observed to all exist at the same time in outer space. A possible way this could happen is outer space is a physical thing of a kind that its density could vary from location to location. Earlier galaxies would form in denser locations and later galaxies would form in less dense locations. Thus, galaxies of different ages could exist at the same time.

Another possible explanation for how galaxies of different ages could exist at the same time is presented in the third paragraph later.

- Observations indicate our universe is expanding. One way this could happen is the outer space is still being brought into being and is thus pushing the tangible things in our universe increasingly further away from each other. The outer space could do such pushing only if it is a physical thing.

- The spirit world might still be bringing our entire universe into being such that not only is more outer space still being brought into being, so are the three physical building materials still being brought into being.

This could also be another possible explanation why very old, moderately old, and very young galaxies exist at the same time. The older ones were formed earlier from building materials brought into being earlier and the moderately old and younger galaxies were formed from building materials brought into being later.

d. The complexity in the designs of living things on Earth, especially the design of us humans, indicates the spirit world must have had a lot of experience designing living things. Since living things emerged on Earth only recently when compared with the age of our universe, it seems unlikely the spirit world gained all its experience in designing living things within the length of time living things existed on Earth.

e. This suggests the spirit world must have been designing living things for a very long time. This could be the case if numerous universes existed in the past and the spirit world had designed living things to reside in each of them. Again, this supports the possibility that numerous universes have existed in the past and more will exist in the future.

2. **Relative extent of evolution among living things on Earth:**

The impressive extend to which many of the living things on Earth have evolved indicates evolution must have been going on for a long time, way before living things appeared on Earth and before our universe came into being. This is especially true for us humans on Earth since we obviously evolved further than any other living thing on Earth and thus had been evolving longer than any other living thing on Earth.

a. A possible way evolution could progress to the extent it has for many of the living things on Earth is that a large number of past universes have come and gone, and evolution in general have spanned over most or all of the numerous past universes.

b. The speculation that universes can end after being brought into being is supported by the fact that various processes are going on in our universe that would eventually bring our universe to an end.

For example, every star will eventually use up all its fuel and then our universe could no longer support physical life. Black holes will continue to swallow up everything and our universe would eventually become just one enormous black hole, and that could be a way our universe could end.

This is assuming that our universe is not continuously being brought into being, but was brought into being once for all by one big event such as a Big Bang.

c. Different living things on Earth appear to have evolved to different extents instead of all having evolved to the same extent. Different living things are bound to be designed and brought into being at different times. As the spirit world constantly grows its ability to design a variety of different living things also constantly grows. Therefore, at any point in time, different living things would have had different lengths of time to evolve and thus they would have evolved to different extents.

The extent of evolution for a given species of living thing would also depend on the diversity of experiences it has gone through. A lot of different experiences would generate a lot of new pieces of knowledge that gets added to the spirit of a living thing. The spirit thus evolves and the evolutionary changes would manifest themselves as changes to the living thing on Earth.

Thus, such a species of living thing will evolve more than a species of living thing that does not go through very many different experiences regardless of when that species was designed and brought into being. Species on Earth that have not evolved much are referred to as living fossils. Examples are lungfishes, horseshoe crabs, and the Latimeria Chalumnae species of fish.

d. Another possibility is that as the spirit world becomes increasingly able to design increasingly complex universes and increasingly intelligent and complex living things, the more complex universes would need a more complex combination of living things to produce environments that would be hospitable for the more highly intelligent living things.

Therefore, the spirit world would design a lot of new living things that could participate in producing such environments. This would include some that are less complex and less intelligent than some that were designed and brought into being much earlier. These new living things would have had less time to evolve than livings that came into existence earlier. We could see how this could be the case for producing the various environments on Earth that are hospitable to us humans.

3. **Evidence of retracing:**

An evolutionary progression would have to take place in the spirit world first before it can be manifested on a living thing in a universe. This is because any evolutionary change that could happen on a living thing has to be done by the spirit world, and the spirit world has to have the spiritual form of the change existing in the spirit world first.

When the spirit world designs and brings a new universe into being, it would have an enormous collection of evolutionary progressions that have already taken place in past universes ready to be used to enable living things to emerge in the new universe through the process of retracing these already existing evolutionary progressions.

According to the spiritual model presented in Reference 1, contrary to the beliefs of some individuals, the spirit world does not do "sudden creations". In the case of our physical universe, it does retracing of existing evolutionary progressions, starting with some very primitive physical forms of living things such as slimes, and through retracing it enables the primitive physical forms to fairly quickly progress to becoming fairly advanced physical living things. The process goes so quickly that there is little or no chance for "missing link" fossils to be created.

Therefore, the fact that no fossils of missing links have been created for numerous living things, including the human species, indicate retracing has taken place. Some individuals perceive the absence of fossils of missing links as being evidence of sudden creation.

According to the spiritual model presented in Reference 1, there is more to evolution than suggested by conventional notions. For example:

4. Retracing is a concept separate from conventional notions about evolution. Its existence is revealed by the spiritual model presented in Reference 1, and by a lack of fossils of missing links on Earth.

5. A lack of fossils of missing links could also indicate numerous past universes have come and gone and that evolution

progressions span over past universes.

More about evolution and retracing is presented later in this chapter.

11.2. Two Kinds of Evolutionary Progressions Exist

This section of this chapter pertains mainly to our particular physical universe, especially when we are talking about evolutionary progressions of the second kind. What is said would be different for each universe. For simplicity, the discussion will be applicable mainly to our particular physical universe.

Two kinds of evolutionary progressions could take place. One is well known, and it is often referred to as being "survival of the fittest". It is referred to here as being "evolutionary progressions of the second kind". The other one is revealed by the spiritual model presented in Reference 1. It is referred to here as being "evolutionary progressions of the first kind". According to the spiritual model in Reference 1, the spirit world is constantly evolving, and the "evolutionary progressions of the first kind" is directly associated with the continuing evolution of the spirit world. That is why I consider it "the first kind" whereas the other kind refers more to physical evolutionary changes of living things residing on Earth.

1. **Evolutionary progressions of the first kind:**

    The first kind is produced by the spirit world being constantly growing as new pieces of knowledge are constantly being generated and added to it. This produces a continuous increase in; intelligence, various mental abilities, and all other abilities for the spirit world. Therefore, it includes a variety of continuing evolutionary progressions in the spirit world. These evolutionary progressions being a part of the spirit world are thus available for spirits of living things to gain access to them in small bits that then become a part of the spirits of the living things.

We might ask, why only small bits and not the whole thing? The answer is that any evolutionary process would progress naturally gradually. For example, if a species of sea dwelling creatures were to become a species of land-dwelling creatures, it does not develop legs in just one generation. It takes a very large number of generations to gradually make the change, a bit at a time each generation.

Similarly, an evolutionary process that enables a living thing to become noticeably more intelligent would also not happen in a single generation. Each generation could become a bit more intelligent. An accumulated increase would become more noticeably after a large number of generations.

Slow progressions appear to be consistent with how things work in the spirit world. The spirit world grows gradually and evolves gradually as new pieces of knowledge are gradually generated and gradually added to the spirit world. The spirit world does not have instantaneous abrupt increases in its growth or in its evolution.

Similarly, once the spirit world brings a new universe into being, the spirit world gradually gets the universe into a condition that would be hospitable for living things. For example, it took the spirit world billions of years to get our universe hospitable for living things. It was done gradually a bit at a time and not abruptly. Incidentally, this is why earlier in the presentation and in Reference 1, I speculated that our universe was brought into being gradually instead of by a Big Bang, and that our universe might still be in the process of being brought into being.

The general way things appear to work in the spirit world is gradually. Gradations are a major attribute of the spirit world, and they are also gradual. I.e., there are no abrupt changes as being a part of any of the gradations. Since the spirit world has to be extremely wise, this could indicate that doing things smoothly and gradually might be one of the wisest ways of getting things done. We can think of a whole lot of examples of this in our lives.

2. **Evolutionary progressions of the second kind:**

> The second kind is produced by naturally occurring variations in the species of a living thing with each generation together with naturally occurring environmental changes. If an environmental change is large enough it would alter which of the variations of the species would be the most viable in the changed environment. The variation that is the most viable would then reproduce the most. An evolutionary progression of the second kind would thus happen with the species. This process is often referred to as being "survival of the fittest".

"Survival of the fittest" is usually taken as being the mechanism for how evolution works. However, the spiritual model presented in Reference 1 reveals that there is also another mechanism, and it is just as logical and as real as is the "survival of the fittest" mechanism. The spirit world itself is constantly evolving, and this other mechanism is directly associated with the evolutionary progressions constantly ongoing in the spirit world.

This mechanism could explain certain things about how living things evolve that have not been explained before. For example, it could explain how various species of nonhuman living things could become highly intelligent even if their physical appearances might lead us humans to not realize how intelligent they are. In other words, their mental evolution somehow got ahead of their physical evolution, and this other mechanism could explain how this could happen. Such living things include dolphins, elephants, whales, octopi, and certain birds.

The reason we have not been aware of this mechanism before is because we have not been aware of how things work in spirit world until the spiritual model in Reference 1 was formulated. This mechanism produces what is referred to here as being "evolutionary progressions of the first kind".

3. **How evolutionary progressions of the first kind work:**

New pieces of knowledge are constantly being generated and added to the spirit world. The spirit world is thus essentially constantly evolving and becoming increasingly intelligent because of its constantly increasing amount of knowledge. It is also constantly becoming more creative, developing more abilities, and becoming generally more advanced, etc.

Such evolutionary progressions in the spirit world would result in the formation of new spiritual entities and/or in the growth of certain existing spiritual entities. All such spiritual entities are spiritual expressions of the evolutionary progressions. Because every spiritual entity in the spirit world is directly or indirectly a part of one another, the spirit of a living thing could always gain access to these evolutionary progressions by going through certain kinds of experiences.

A living thing that is curious and enjoys learning would more often gain access to these evolutionary progressions. For example, we humans are very curious about everything and we are constantly doing research on a variety of things. Therefore, numerous spirits of our species are constantly exploring and trying to figure out an unknown. These human spirits are thus applying their levels of intelligence intensely.

By applying their levels of intelligence intensely the human spirits are essentially searching for additional sources of intelligence to help with their efforts. This would enable each of them to gain access to a bit of the kind of evolutionary progression that is associated with the kind of research that each of human spirit is doing. For example, a physicist doing research in physics would gain a bit of an evolutionary progression that has to do with the area of physics of the research. A medical doctor doing research on a medical topic would gain a bit of an evolutionary progression that has to do with that area of medicine of the research.

Etc.

This is in accordance with how the spiritual model presented in Reference 1 explains that every experience would generate new pieces of knowledge and/or gain access to various existing pieces of knowledge. In this case the experience that a researcher who is applying his or her level of intelligence intensely would goes through would enable him or her to generate new pieces of knowledge and also gain access to the spiritual entities that are the spiritual expressions of the evolutionary progressions of the spirit world. This would increase the researcher's level of intelligence and other abilities a bit.

In this manner our species would produce experts in every field of expertise. Some experts would also work as professors in universities. This would enable university students to gain access to the latest knowledge in all professional fields.

This could explain how our species has become the most intelligent among all the species of living things on Earth. Any bit of increase accomplished by one generation might not be obvious, but the increase will carry over to the next generation. Therefore, after many generations of increases the total increase will become more obvious.

In my mind, this discussion also explains why the spirit world made mankind's secondary purpose for being here on Earth as being "to generate as many new pieces of knowledge as possible".

Besides helping the spirit world grow, this secondary purpose enables the spirit world to form evolutionary progressions. While much of the new pieces of knowledge are randomly generated, the spirit world would assemble the new pieces of knowledge along with existing pieces of knowledge to form evolutionary progressions. And then it would make the evolutionary progressions available for the spirits of living things to gain access to them.

Accordingly, a species of a living thing could achieve an evolutionary progression only if the evolutionary progression already exists in the spirit world. This also points out how it is that evolution has to take place first in the spirit world before it could take place physically with a species of a living thing. Again, we are talking about our particular physical universe and the living things residing on Earth. This would be consistent with the notion that we could use our creativity to produce something only if its spiritual form has already been created by the spirit world and thus exists in the spirit world.

Mankind's primary purpose for being here on Earth is to apply in positive and constructive ways what we learn about our physical world so as to make life on Earth to be as close as possible to be how life is in the spirit world. The way the spirit world goes about forming its evolutionary progressions and then making them available for all spirits of living things to gain access to them is a good example of one of the ways we humans ought to behave in order to fulfill mankind's primary purpose for being here on Earth. For example, we should not be greedy by trying to possess and/or control everything we could, but instead we should try to make a contribution to mankind in whatever ways we could.

4. How evolutionary progressions of the second kind work:

>Variations of all kinds are always produced by living things each time new members are produced. The variations are usually so slight they are not obvious. If the environment happens to change enough for any reason, then the variation that would be the most viable in the changed environment would be different from the variation that used to be the most viable in the preceding environment.

>The variation that is now the most viable would then reproduce the most. The species would then have undergone an evolutionary progression of the second kind.

At first the evolutionary change would not be noticed because the number of the new variation that is the most viable would still be few compared with the total. After numerous reproductions, the number of the new variation would become large enough such that the evolutionary change would become more obvious. Eventually the variations that used to be the most viable would go extinct. This process is often referred to as being "survival of the fittest".

Usually, when an evolutionary progression occurs it is a combination of both the first and second kinds. This is because for the species of a living thing to adapt to any large change in its environment, it usually needs to use many if its abilities intensely to do so, especially its mental abilities. This is why an evolutionary progression of the second kind tends to always be accompanied with some increase in intelligence; i.e., with some amount of evolutionary progression of the first kind.

Living things that go through the largest number of diverse experiences would evolve the fastest and the most. Humans on Earth would be such a species among all species on Earth. Conversely, living things that have not gone through a whole lot of diverse experiences for thousands of years would hardly change for thousands of years. They are often referred to as living fossils.

It is possible for evolutionary progressions of the first kind to take place for the spirit of a living thing whether the spirit is enabling a living thing to exist in a universe or not. This means the spiritual entity that would serve as the spirit of a living thing could continue to become more advanced, become more intelligent, have increased mental abilities even when it is not serving as the spirit that would be enabling a living thing to exist in a universe. Although this does not necessarily always happen.

On the other hand, evolutionary progressions of the second kind could not take place unless the spirit of a living thing is enabling a living thing to exist in a universe. This is because the "survival of the fittest" is only a concept and not a practice in the spirit world. It could be a practice only in a universe or something similar that is designed and brought into being by the spirit world. Also, as revealed earlier in the presentation in this book, in order for a universe to be able to produce the kind of new pieces of knowledge that would help the spirit world restore balance, "survival of the fittest" would likely have to be a practice in a universe.

This could explain why some nonhuman living things on Earth are very intelligent even through they are not humans. The spiritual entities that would serve as their spirits have gained access to some of the evolutionary progressions that are in the spirit world while they are not serving as the spirits of living things and enabling those living things to exist in a universe. Such nonhuman living things would include dolphins, whales, elephants, octopi, and certain birds.

This could also explain why certain animals could be domesticated by humans easier than others, and why some animals seem to naturally like humans. It is likely because the spiritual entities that would serve as their spirits have also gained access to some of such evolutionary progressions while not serving as spirits for those living things and thus not enabling living things to exist in a universe.

In case we might be wondering about how a living thing would look like in one universe compared with how it looked like in another universe, a living thing is not likely to look the same in every universe. Each universe is likely to be dimensionally different, might be nonphysical whereas the preceding one was physical and vice versa, would be made of different building materials, etc. All such differences are bound to mean the way a living thing will look like will be different in each universe.

11.3. The Process of Retracing

The meaning of the word "retracing" as it relates to the evolution of living things is first used in Reference 1. The process of retracing appears to be real and in action as revealed by the spiritual model

presented in Reference 1. It could explain in a logical and explicit manner why the fact that missing links have not been found for many living things such as us humans, dolphins, whales, wooly mammoths, etc.

This indicates retracing is applied by the spirit world to quickly bring into being in a new universe the advanced form of a species of living thing that has already evolved to their advanced stages when they resided in past universes. As explained in Reference 1, the spirit and soul of a living thing contain the entire history of the experiences the living thing has gone through. Therefore, its evolutionary path leading up its advanced stage exists in its spirit and soul. This means the living thing could retrace this evolutionary path to quickly arrive once more at its advanced stage in the new universe in which it now resides. And then its evolutionary progressions could resume in the new universe where it left off in the preceding universe in which it resided.

We might ask; why would retracing as described be necessary? Why couldn't a species simply continue evolving starting from where it left off in the last universe?

The answer is the spirit world has grown and changed a lot since the last time it designed and brought a new universe into being. Therefore, when it needs to design and bring a new universe into being, the new one would be completely different from any of the past. It would be more advanced, bound to be made of different building materials, likely to be dimensionally different, might be nonphysical whereas the preceding one was physical or vice versa, etc.

This means before the spirit world could bring any species of a living thing into being in the new universe it has to first get familiar with how to work with the new building materials the new universe is to be made of, even though the spirit world knows what kind of building materials is needed to form the kind of universe it needs.

By analogy, suppose we need to bake a special kind of cake that we never baked before and have never worked with the special ingredients before even though we know what ingredients to get. Therefore, the first thing we would do is to get acquainted with the ingredients to see what they are like and what happens when they are baked, etc. We could start by baking several simple things such as biscuits and cookies. We would gradually be ready to bake that special kind of cake.

Then suppose we next have to bake a different kind of special cake, again using ingredients we never used before. We would again have to first get acquainted with the ingredients. But this time we could apply what we previously learned and only need to bake one or two simply things to be ready to bake the cake. The retracing of our experiences enabled us to more quickly be ready to bake the cake this time.

An example of how the spirit world would use retracing to bring a species of living things into being in a new universe relatively quickly is as follows:

1. The spirit world could first only bring into being the building materials the new universe is to be made of.

2. The spirit world would then get acquainted with the building materials to see how they feel and what they could do. In the case of our universe, this getting acquainted period included forming: electrons, protons, and neutrons, the various atomic elements, galaxies and stars, various kinds of molecules, etc.

3. When the spirit world is ready to bring living things into being in the new universe the forms of the living things has to be such that they could be formed from items such as those mentioned in the preceding paragraph. This means the forms has to be the most primitive possible.

4. Once these primitive forms are brought into being, the spirit world could start applying its retracing process to fairly quickly step by step and generation by generation gradually progress to the advanced stages of the living things. This could still take

perhaps thousands of years, but would be much faster than millions of years.

5. The primitive forms would still need to go through many of the more advanced forms such that it could look like a very fast version of evolution. The process might be able to bypass some of the less crucial steps, and this too would make the retracing process go faster.

6. The relatively fast progression would not appear fast to a human observer because it is still slow compared with the typical human attention span. But it would be fast enough such that fossils would not be left behind for every state the species has gone through during the retracing.

7. Upon arriving at its advanced state on Earth, the species could then resume its evolutionary progressions starting from where the species left off from the last universe in which the species resided.

11.4. Incompatibilities Could Form Among Living Things

Have we ever wondered why living things on Earth have allergies and other incompatibilities? If all living things on Earth had evolved together, we would think incompatibilities such as allergies would not develop. Therefore, we might ask; how and why do incompatibilities happen?

As revealed by the spiritual model in Reference 1, it is unlikely every species of living thing existing in the spirit world would be selected to reside all together in any one universe. Some would be selected to reside in one universe; some others would be selected to reside in another, etc. such that the species are not going to evolve all together. Thus, the evolutionary progressions are more likely to be different for each species.

Consequently, incompatibilities are bound to develop.

This appears to be the most likely reason why incompatibilities developed among various species.

Another possibility is that incompatibilities are simply one of many possible forms of imperfections. The spirit world will always be correct but incomplete, and therefore never perfect. This means the living and nonliving things it creates could never be perfect.

However, I think the first reason is more likely the cause of incompatibilities. I think imperfections are less likely to cause incompatibilities than different evolutionary paths would.

11.5. Humans Are Likely to Have Been Evolving Only Moderately Long

New species of living things are constantly being created from scratch by the spirit world. The newer they are the less time they have had to evolve, and therefore they are more likely to be perceived as being more primitive at any given point in time. Conversely, the older the species of living thing the more time they have had to evolve, and thus the more advanced they could be.

In Reference 1, I speculated that UFOs and UAPs visiting Earth indicates that species of living things exist that are more advanced than we humans. Also, we humans have made a mess on Earth, and this also indicates we could be a lot more advanced that we currently are. These are the main reasons I think we humans are not among the oldest living things in the spirit world, because it seems we have been evolving only a moderately long time.

We might then ask; why are we humans selected by the spirit world to be the most intelligent living things to reside on Earth? We have made a mess on Earth. Therefore, the selection of us doesn't seem to be a very good one.

The possible answer is as follows, and it is not intended to excuse humans for being stupid but is to explain why we are selected particularly to reside in our particular universe:

1. As explained earlier, the spirit world would design and bring into being a new universe as one of the ways to help restore balance in its state of knowledge.

2. Each new universe would be totally unfamiliar to any species of living thing in the spirit world. Therefore, it is important that the most intelligent species to reside in a new universe be one of those that could think for themselves in order for it to be able to generate the kinds of pieces of knowledge that would help restore balance for the spirit world. This could not be accomplished if all the species were to require the spirit world do all the thinking for them. Otherwise new universes would not be needed to help restore balance.

3. In the case of our universe, the spirit world needs to have multiple locations where living things could reside. While our universe is likely to fulfill the same primary purpose for every location, the spirit world most likely wanted to try numerous approaches to fulfilling this purpose. Perhaps it is because fulfilling this particular primary purpose is particularly tricky, and it wanted "redundancy" to help assure successful fulfillment of the purpose somewhere in our universe.

4. Our having made a mess on Earth could be evident that fulfilling this primary purpose is tricky. This could mean we humans are not among the most advanced species of living things that exist in the spirit world. And, the fact that visitations by UFOs and UAPs on Earth could further indicate this is the case. In other words, the living things that are able to visit Earth as UFO's or UAP's are more advanced than we humans are.

5. What all this could mean is that we humans is one of those

species that show a lot of promise, and that we were selected to reside on Earth to be the most intelligent living things on Earth to give us a chance to see what we could do and how much further we could advance. If this is the case, then we humans need to very soon "wake up" and start doing better.

6. So far, we have not done very well. But we still have time, assuming we do not make an even bigger mess than we have al ready made. If we continue to contribute to climate change we just might run out of time because we are likely to be mired in battles over inhabitable land, clean water, food supply from land and seas, loss of rain forests and thus loss of breathable air, etc. The battles could lead to the use of the kinds of weaponry that would accelerate the shortages of everything mentioned. Our chances of proving we could do better would be gone as we will be become mired increasingly deeper in battles.

7. At the end we humans might only manage to succeed in proving that we are not among the most advanced species of living things in the spirit world.

8. However, keep in mind that we still have time, at least as of this moment.

Chapter Twelve

An Analogy Between Evolution and Maturation

12.1. Maturation

We humans begin life as babies and could hardly do anything. Our parents do everything for us including thinking for us. They also pretty much limit what we are allowed to do mainly to assure our safety. We begin learning things right away.

We eventually begin to craw and then walk, express ourselves, feed ourselves, and play with toys and with others. We continue to learn things, and we start to think for ourselves. We begin to have a lot of freedom to explore.

Soon we are able to think completely for ourselves, do most things ourselves, solve problems, etc. We start to learn about the more complex things about life on our own, in school, in college, and in life. We start do things with others as a team. We start to have a lot of independence and responsibilities. We start to form wisdom.

As an adult we take on a lot of responsibilities. We need a lot of wisdom and we are constantly in need of more. Our status in every part of our life becomes important to us; e.g., in our family, our workplace, our community, etc. And for a while we think this is what life is all about. As we age, we start to think about what our purpose is in life and perhaps also about what our species' purpose is for being here on Earth.

Our life on Earth will eventually end and so will our maturation process on Earth.

12.2. An Analogy Exists Between Our Species Evolutionary Process and Our Personal Maturation Process

Our species' existence began in some primitive form that spirit world is able to form out of the building materials that make up our universe. The spirit world does everything for us, including thinking for us. We are very limited in what we could do. We begin to evolve right away although very slowly.

We eventually became able to move from one location to another. We start to explore and to do some of our own thinking.

Soon we became very mobile creatures and can do quite of lot of our own thinking. We start to use our mental abilities to enhance the viability of our species. We start to form tribes and shelters to further enhance the viability of our species. A leader is needed to maintain cohesiveness in a tribe. We collectively learn how to make tools, other equipment, and weapons.

Our species eventually evolved to become highly intelligent and are able to handle a lot of responsibilities for ourselves, our families, and our communities. Personal status becomes more important in every part of our lives. Various forms of spirituality begin to develop to give more meaning for life.

Our species' existence on Earth will eventually end and so would our species' evolutionary process on Earth.

12.3. Learning from Our Maturation Process and Shaping Our Species Evolutionary Path

The idea of learning what we could from our personal maturation process and then applying that knowledge to shape our species' evolutionary path makes a lot of sense when we consider what mankind's primary purpose is for being here on Earth, as revealed by the spiritual model presented in Reference 1. As stated earlier in Chapter One, Section 1.7. and Chapter Seven, Section 7.2., on how to fulfill mankind's primary purpose is as follows:

Mankind's primary purpose for being here on Earth is to gain as much knowledge as possible about our physical world and the spirit world, and to apply that knowledge to make mankind's life on Earth to resemble as close as possible to how life is in the spirit world.

This statement reveals why we humans have not been able to make much progress in helping our species fulfill its primary purpose. It is that before the spiritual model in Reference 1 was formulated we did not have any idea how life is in the spirit world.

One thing we humans could do as the most intelligent living things on Earth is to purposely shape our species' future evolutionary path. We cannot do anything about the past path our species' evolutionary process has gone through but we could shape our species' future evolutionary path.

The alternative is to simply let it be what it could be by chance, which is what we seem to have been doing. And, so far this has not gotten our species to progress further toward fulfilling its primary purpose as evident by the mess we humans made on Earth. As far as we can tell, no other living thing on Earth is able to shape its species' future evolutionary path. We seem likely to be the only living things on Earth that could do that, and there is very likely to be a reason for it.

We are likely also to be the only living things on Earth to do the following:

1. At some point as we mature from being helpless babies to being responsible adults, most of us are likely to suspect each of us has

a purpose in life, different for each us, and each of us is to figure out what it is and how to fulfill it.

2. In addition, most of us are likely to also suspect mankind as a whole has a purpose for being here on Earth.

3. This suggests each of us has two purposes, one for each of us personally, and one for mankind as a whole. We might say one is local and the other is global, or one is for family and the other is for mankind. And, each of us is to figure out what they are and how to fulfill them.

Figuring out our personal purpose and how to fulfill it is bound to be easier than figuring out mankind's primary purpose for being here on Earth and how to fulfill it. Part of the reason is because evolutionary processes move much slower than maturation processes. Therefore, we tend to not think about mankind's primary purpose for being here on Earth or how to go about fulfilling it. However, the spiritual model presented in Reference 1 revealed two things that would help us with both.

1. The spiritual model in Reference 1 was able to reveal what the primary purpose is for mankind to be here on Earth and what needs to be accomplished to fulfill it. These are as stated earlier in Chapter One, Section 1.7. and Chapter Seven, Section 7.2.

2. The spiritual model also reveals an analogy exists between our personal maturation process and mankind's evolutionary process. This analog is likely to be put in place by the spirit world to help us figure out how we could go about helping mankind fulfill its primary purpose for being here on Earth.

This analog is valid and real because both processes involve life altering experiences. According to the spiritual model in Reference 1, an experience generates new pieces of knowledge that get added to the spirit world and/or gains access to some existing pieces of knowledge. And, the kinds of pieces of knowledge involved in both processes would be the kind that would result in maturation and/or evolutionary changes.

This means that what we could learn about our personal maturation process could be applied to help us figure out what we could do to shape the future evolutionary path of mankind's evolutionary process such that mankind would always be progressing toward fulfilling its primary purpose for being here on Earth.

How this could work is as follows:

1. Maturation is a personal and individualized process. It goes from its beginning to its end within the lifetime of an individual and it takes place in just one universe, at least in the case of our universe. The maturation process moves along at a pace that is fast enough to enable us to generally understand what is going on and to figure out what it is we must generally do keep our maturation process moving along in a normal manner.

2. Evolutionary processes happen with species of living things. They are constantly ongoing and never end as they span over multiple universes. They consist of a first kind and a second kind. Their pace is so slow that we tend to not think about them, let alone trying to figure out how to shape their future paths or why we should even bother to do so.

3. The first kind has to do with mental abilities and various other abilities, and it is constantly ongoing without interruption.

4. The second kind progresses in segments with each segment starting and ending within the lifespan of each existing or past universe in which the human species is residing or has resided. The present segment is going on in our particular physical universe, and it is involved mainly with physical changes. The segments that had gone on in past universes would have been involved with other kinds of changes. For example, for a past universe that happens to be nonphysical, the segment would

have been involved with some nonphysical changes.

5. The spiritual model presented in Reference 1 revealed what the primary purpose is for our species for being here on Earth, and it also revealed an analogy exists between our personal maturation process and mankind's evolutionary process.

6. This analogy is likely to be put in place by the spirit world and is thus available to help us figure out what to do to shape the future evolutionary path of mankind's evolutionary process such that mankind would always be progressing toward fulfilling its primary purpose for being here on Earth.

7. **One of the most significant things we learn in our maturation process is that somewhere along the way we need to work together with others as a team or as partners in order to make accomplishments that have meaning for mankind. By analogy, somewhere along mankind's evolutionary process we need to work with one another as a team or as partners in order to help mankind do the things it need to do the fulfill its primary purpose for being here on Earth.**

8. For us humans to collectively take an active role in shaping the future evolutionary path of mankind's evolutionary process in a positively and constructively manner would be consistent with mankind's primary purpose.

   This is because we humans would be acting in a manner that would be as close as possible for us to be as if we were actually a part of each other. In the spirit world our spirits are actually a part of each other. Therefore, such a collective effort would be meaningful and it would be an effective way of making mankind's life on Earth resemble as close as possible to how life is in the spirit world.

I think one of the reasons we humans have made a mess on Earth is because we somehow lost sight of the fact that we need to take care of mankind as well as take care of ourselves, our families, our community, and our civilization. We need to think longer-term and longer-range. Life is more than just the life we carry out on Earth, and each of us is more than just who we are individually.

I think if we could understand how it is that our maturation process is an analog of our species' evolutionary process, we might realize our evolutionary process is as important to mankind as our maturation process is to each of us personally. Both are doing the same kind of thing, except the evolutionary process is much slower, is on a much larger scale, and it is much more significant in the long run.

Chapter Thirteen
Hang-Ups

13.1. General

This chapter touches upon things that are only concepts and not practices in the spirit world. Hang-ups are among such things and they are the focus in this chapter, because in my opinion they are one of the major causes of poor overall human behavior.

13.2. Hang-ups Could Severely Influence a Person's Behavior

We all have hang-ups. They come in all kinds and all sizes such as regarding; foods, places, events, germs, holidays, people, etc. They are personal and easy to develop. They vary in their amount of influence on our behavior. Most are minor and usually not serious. Some are so severe they color everything we do. When we have one of those, we will be amazed by the power it holds over us. The most severe kinds usually involve people.

A severe hang-up is like a bad itch. We have to scratch it whenever we have a chance even if we know scratching it will not make it better or go away. It doesn't seem to matter whether the situation is appropriate or not, we will scratch anyway. Scratching takes top priority over everything else.

I have seen several cases of severe hang-ups in close proximity. Some would last the entire lifetime of an individual. Sometimes it consists of multiple bad ones, each adding on top of another to produce one big severe hang-up.

I had a severe one at one time. I was lucky enough to break its grip early enough in life such that I had time to recover without a lot of bad aftereffects. Breaking it was not easy, not quick, and it was painful for everyone involved. As bad as breaking it was, there was a positive side afterwards. Upon being free I could just feel myself grow inside. It was as if my inner growth potential had been confined, dormant, out of sight, out of mind. And, now it was awakened. It felt more liberating than I could ever imagine. Gradually I became a better person, more generous, more caring of others, and focusing less on myself. It is all because my mind is no longer preoccupied with myself and my need to scratch.

To illustrate how powerful a severe hang-up could be, in one such case the severe anger that came with it caused a terminal cancer to develop that resulted in the early death of the individual. For various reasons, that individual wasn't willing to try resolving the hang-up with the people involved, nor was the individual willing to take those people put of that individual's life. The anger distorted just about everything that individual does. For example, that individual would know the proper way to accomplish a goal and could even say what the proper way is. But the individual would not actually do it that way but would instead do what amounts to being the opposite and thus fail to achieve that goal. And, it doesn't seem to matter to that individual that the goal was not achieved. Venting the anger took top priority over achieving the goal, and doing the opposite was a way to vent the anger.

The individual's spouse developed a heart condition because the spouse was absorbing all the anger. Also, as a likely result, the spouse came down with cancer three years before the individual came down with cancer. After the individual's death the spouse's heart condition gradually resolved itself and the cancer was able to be treated such that it went into remission. It is staying in remission so far even though it is a kind of cancer that has an 80% chance of recurring.

The cause of the severe hang-up also caused the individual to develop a dual personality. One personality was normal and other was very angry. The spouse fell in love with and married the normal one. The angry one revealed itself soon afterwards and was present a majority of the time thereafter.

The spouse never stopped loving the normal one as the spouse devoted seven months caring for the individual 24/7 as the individual slowly died of cancer. The stress from caring for the individual caused the spouse's own cancer to worsen, and it became twice as bad as before. The spouse chose to ignore it. After all, it didn't matter to the spouse at the time if it becomes terminal such that the spouse would also die along with the individual's death.

Shortly before dying the individual described to the spouse that the individual could feel two persons residing inside that individual. That was no surprise to the spouse, but it was the first time a dual personality was acknowledged by the individual. The individual then said to the spouse that the individual loves the spouse more and more. That was something the spouse longed to hear for almost half a century, and the spouse finally knew how much the spouse was loved by the individual just before the individual died.

13.3. Hang-ups Could Disrupt Mankind's Efforts to Fulfill Its Primary Purpose

I remembered as a child attending church the pastor would frequently instruct us to not do "worldly things". I was too young to know exactly what that meant except for the few examples he gave such as: don't see movies, don't attend dances, girls should not wear lipstick, etc. Such examples were so incompatible with common everyday observations and experiences that his message was unable to come across very meaningfully, especially for a young child. However, more than seventy years later, I am able to interpret his message more meaningfully. Better extremely late than never, I guess.

We could choose to carry out our life on Earth for the sake of our life on Earth, and thus do "worldly things," or we could choose to carry out our life on Earth for the sake of our everlasting life afterwards in the spirit world. It makes a difference on our values and priorities. If we focus too much on values that have meaning for our life on Earth but have no meaning for our life in the spirit world then we are likely to develop numerous hang-ups that have meaning for our life on Earth but have no meaning for our life in the spirit world.

Since every human spirit is directly a part of every other human spirit in the spirit world and every spiritual entity is directly or indirectly a part of every other spiritual entity, there are no hang-ups in the spirit world. Hang-ups are only concepts, and are not practices in the spirit world.

This could explain why hang-ups could develop very easily on Earth. It is because living things happened to be separate entities on Earth.

Since there are no hang-ups in the spirit world, the spirit world is likely to have addressed the concept of hang-ups, but have not learned how to address hang-ups as practices. But, in order to maintain a reasonable state of balance in its state of knowledge the spirit world needs to learn how to address things as practices as well as concepts, and this includes how to address hang-ups as practices.

This means among the things mankind on Earth is to help the spirit world learn about, learning how to handle hang-ups would be one of them. This could explain why living things are designed to be separate entities in our particular physical universe. It is to enable certain things such as hang ups that are only concepts in the spirit world to be practices in our physical world.

More specifically, we humans are here on Earth to help the spirit world learn how to handle various specific issues, and we are to do it in a manner that is relevant and compatible with how life is in the spirit world. In other words, we are to learn how to do them in a wise, empathetic, compassionate, and constructive manner. In my mind, this is especially true regarding how to handle hang-ups, because I see hang-ups as one of the most powerful causes of human poor behavior.

We humans would thus succeed in accomplishing all such learning for the spirit world when we generate the needed knowledge in a manner that makes mankind's life on Earth be as close as possible to be how life is in the spirit world. In other words among all the experiences we need to go through on Earth to generate the needed new pieces of knowledge specifically to restore balance, our universe is likely to be designs such that those experiences would automatically include the ones that would make mankind's life on Earth be as close as possible to be how life is in the spirit world.

I think this is important to point out because our potential for being able to make mankind's life on Earth to be as close as possible to be how life is in the spirit world could be an added inspiration and motivation for mankind to work toward fulfilling its primary purpose for being here on Earth.

In addition, accomplishing this is not easy when dealing with severe hang-ups. Therefore, having multiple sources of inspiration and motivation would be helpful. This is because severe hang-ups tend to demand top priority, and this tends to keep mankind preoccupied with its part of life that is taking place on Earth instead of paying sufficient attention to its part of life that is taking place in the spirit world. This means we humans need to first resolve, learn how to work around, or get rid of our severe hang-ups, especially the ones to do with people.

13.4. We Don't Have to Get Rid of Our Possessions

In some religions, we humans are supposed to get rid of our "worldly wealth" before we can "enter the kingdom of God". While I could understand why such a concept is promoted, I think it is unrealistic in today's world. Situations have changed a lot since this traditional concept was formulated. But we humans now need to have some level of "worldly wealth" in order to survive and to have the time and opportunity to be productive and contributory to mankind. Otherwise, we could develop severe hang-ups regarding various needs that require some amount of wealth to maintain such as:

1. Where would the next meal come from?

2. How could we maintain shelter?

3. How could we stay clean and have clean clothing?

4. What if we develop a medical problem?

5. What if we have an accident?

6. How are we going to survive when we get old, etc.

While such a traditional concept might have worked in the past it is likely to be outdated and incompatible with present conditions. For example, we are dealing with the following conditions in today's world:

1. The human population keeps growing and is now at a level in which Earth's natural resources are all stretched to their limits and are barely able to support the population as is.

2. Homelessness has become a significant problem, and workable long-term long-range solutions are yet to be found.

3. A lot of families are able to avoid becoming homeless by being very carefully about their finances as long as no emergency or disaster happens. They are not likely to get rid their worldly wealth. Otherwise they would become homeless.

4. A lot of families are unable to provide adequate food for

themselves. They are depending on school lunch programs to help feed their children.

5. Rain forests are being destroyed to convert more land for food production. Rain forests are among the vital "lungs" of the world. Therefore, destroying them would result in a worsening future for mankind and all other living things.

6. A lot more education and/or training are required to hold good paying jobs. Jobs that do not require much education or training usually do not pay much. They and also some good paying jobs are increasingly being taken over by automation.

7. Workers today need to be reeducated or retrained for a new field of work an average of three times over his or her lifetime.

8. The high cost of a college education makes such education, reeducation, or retraining not possible for some individuals.

9. The disparity in pay among the various jobs and positions is ridiculously high and getting worse. In my opinion this makes our nation not as strong or as wealthy as the total dollars would indicate because a lot of it is locked up with the very rich leaving not as much as we think to be in circulation to maintain the nation's strength and health.

10. When social media first came into being, I thought it was wonderful because it could promote communication and mutual understanding among diverse groups of people throughout the world. But social media turned out to increase isolation and loneliness, and decrease empathy. Stanford University Professor of psychology Jamil Zaki described this in his book, Reference 8; i.e., that bad applications of social media have created a lot of complications because of the time and effort required to deal with the problems they cause.

For such reasons I think some of the traditional messages that were formulated in the past need to be updated such they would be more relevant in today's world. The world has changed so much that there is hardly anything from the past that could get by without some updating. Church attendance has been declining a lot. This is a clear indication that some of the messages provided by the various religions need to be updated as well.

In addition, for those of us who live in a nation run by a democratic form of government, the idea of entering into a kingdom of any kind might not resonate with some individuals, perhaps particularly with the young. Also, according to the spiritual model presented in Reference 1, there is no kingdom in the spirit world.

The spirit world is simply the largest spiritual entity that exists at any point in time and it is made up of all the smaller spiritual entities that exist at that point in time. It is not a spiritual entity that is separate from all the smaller spiritual entities such that it would "rule" over them as a king would.

All the spiritual entities collectively make up the spirit world, they are all directly or indirectly a part of one another, and thus they all simply do everything together. This suggests that the way things get done in our afterlife is more like a democratic way of doing things than like a kingdom way of doing things.

In addition, since every living thing is directly or indirectly a part of every other living thing in the spirit world, "having an ego" is only a concept and not a practice in the spirit world. This means there is no living thing in the spirit world that has a big ego or would consider itself to be a supreme being who needs to be worshiped. It thus follows that the idea that mankind needs to worship a supreme being is only a concept and not a practice in the spirit world, according to the spiritual model in Reference 1.

This is not to imply God does not exist. The spirit world is inclusive of all concepts of God and all religions, traditions, customs, ways of life, etc. as explained in Reference 1. However, the concept of needing to worship God has to be only a concept and not a practice in the spirit world.

Also, the analogy between maturation and evolution is applicable here. For example, parents would normally not expect or want a child to worship them forever. A child growing up would probably worship his or her parents at an early stage of their maturation in which he or she depends a lot on the parents for guidance. Therefore, the stage of worshiping the parents could exist, but it would last only briefly. And then the child would be pursuing his or her independence. Parents would normally likewise like to see their child pursue his or her independence after the child has matured pass a certain stage of maturation.

By analogy, I think God would similarly like to see us humans evolve beyond the stage in which we worship God to then earnestly start thinking for ourselves in positive and constructive ways. I think God would not want us humans to worship him or her forever. Instead God would expect our stage of worshiping him or her to last only briefly just as how the stage of a child worshiping his or her parents would last only briefly.

It thus follows that we should not think worshiping God would be good enough in terms of fulfilling our primary purpose for being here on Earth. According to the spiritual model in Reference 1, God needs us to actively and purposely care for him or her as well as our receiving care from him or her. It has to be an active and purposeful two-way street. Therefore, God would want to see us make the natural partnership we have with him or her (or with the spirit world) be mutually beneficial, as explained in Reference 1.

Therefore, in my mind, instead of continuing to encourage people to worship God, it is time to go beyond that stage and to start doing something that is more active, more purposeful, and more constructive. For example, we need to start behaving as if everyone is a part of one another, just as how every human spirit is a part of one another in the spirit world.

Actually, if we think about it, a kingdom of any kind and the need to worship a supreme being are notions that seem to be more worldly than spiritual since kingdoms do not exist in the spirit world and worshiping a supreme being is not a practice in the spirit world, according to the spiritual model. Therefore, a religion that promotes such notions but yet discourages worldly behaviors would be inconsistent with its messages and thus could be counterproductive in its teachings.

I think it is more inspiring and more motivating to encourage everyone to consider themselves as having one more child than the number of children they actually have. And, that the extra child would be mankind. For example, I have two children, but in my mind, I have a third child. I consider mankind as being my third child and I would care for it in a manner very much like how I would care for my two actual children.

Then we might ask, while we might be able to do this monetarily, but how could we do it in other ways? After all, mankind is such an enormous entity and God or the spirit world is nonphysical such that we cannot sense them with our major senses.

This is a significant question and the answer is as follows:

It is important to care for God or the spirit world in other ways. I say we need to give care to them just as we would receive care from them. And, we could care for them by physically, empathically, and compassionately caring for their representatives who are here with us on Earth. Their representatives would include most importantly every other human being besides ourselves on Earth. Each of us exists on Earth because some small portion of God or the spirit world is enabling us to exist on Earth. Accordingly, each of us could be perceived as being a representative of God or the spirit world.

Some examples of how to give care to them would be the following

Be kind, empathic, and compassionate with others. If we see someone needing help and we are able and available to give help, we could go ahead and give help. Behave in a partnership manner with others when possible. Whether it is with our medical doctor, our car repair person, our house painter, or simply a neighbor, interact with them in a partnership manner when we can. Realize they mean something to us and let them realize they mean something to us. When we are having some worker do some work around the house, do what we can to make what they are doing easier for them and more effective for them and us. In other words, think of them as well as ourselves instead of thinking mainly of ourselves. By our doing that, we would find things go smoother and more satisfyingly every time.

13.5. What Kind of Person Would Be Good to Work for?

A close friend once asked me what kind of person would be a good person to work for? I said look for someone who is secure. I believe the more secure a person is, the less likely that person is to develop severe hang-ups. I would also add, look for someone who is very creative. I believe a someone who is both secure and creative is more able to resolve problems in an effective and positive manner. He or she is less likely to have severe hang-ups that could take highest priority in his or her thinking and actions.

On the other hand, a person who is insecure is more likely to first seek blame and would not acknowledge he has any part of a problem. If he or she is also not creative, he or she would also be less likely to come up with an effective and positive resolution. He or she is more likely to focus first on self-preservation, because he or she is more likely to have a hang-up such as being afraid of losing his job or of having someone below him rising up to take over his job. He or she is likely to feel insecure in his or her position perhaps because he or she is likely not creative enough to do a good job of fulfilling the responsibilities of his or her position. Thus, self-preservation takes top priority.

Signs of this would include blaming others if things don't go well, taking credit when things go well even if undeserved or not fully deserved, exaggerating or spinning facts to make things sound way better for himself or herself than reality, or doing similarly to make things sound way worse for others than reality, etc.

These same kinds of considerations would be relevant also for the following questions:

1. What kind of person would be good to vote for in elections?

2. What kind of person would make a good partner?

3. What kind of person should we choose as role models?

4. Etc.

Chapter Fourteen
Our Choice of Dominant Basis for How We Carry Out Our Lives

Up to now, we humans have been pursuing mainly technical advancements and therefore we have learned a lot about how thing work in our physical world and how current life is on Earth. We have not made much progress pursuing spiritual advancements and therefore we have not learned much about how things work in the spirit world or how life is in the spirit world.

We might ask: why do I say "current life" instead of "life"? The answer is because the way we carry out our lives could change. In fact, in order for mankind to fulfill its primary purpose for being here on Earth, mankind's overall behavior has to improve and therefore the way we carry out our lives has to change accordingly.

Based on what we know about how current life is on Earth, the dominant basis we have chosen for how we carry out our lives on Earth is as follows:

14.1. Ego and the Desire to Enhance Survivability

This choice might seem reasonable since the major attributes about mankind's current life on Earth are as follows:

1. Every individual on Earth is a separate entity. It is likely because we are separate entities that all of us either developed or naturally have an ego ranging in size from very small to very big.

2. The materials and energies making up our physical world could be possessed and/or controlled by individuals.

3. Living things need to consume other living or formerly living things to survive.

4. Every individual would naturally like to enhance his or her survivability by maintaining some level of safety margin beyond what is absolutely necessary for survival.

5. We tend to admire those individuals who managed to achieve a high level of safety margin by, for example, becoming very rich, and we want to do the same.

6. Such individuals most often achieve such status by finding ways to possess and/or control a lot of materials and energies that make up our physical world.

7. But if this method is not possible for some individuals to pursue, then those individuals would try to achieve status in numerous other ways, good and bad

    Examples of a good way would be to do volunteer work to help others in need of help, and some people would choose a profession that would contribute positively and constructively to the future of mankind.

    Examples of a bad way would be some individuals would find bad ways to apply our technical advancements and thus feel great about messing up other people's live, and some individuals might feel great supporting various powerful people who do things that messes up people's lives.

These attributes combined with our current state of human nature would result in most individuals developing hang-ups regarding their status in life. Those individuals with very big egos would have bigger hang-ups. A big ego plus big hang-ups could often lead to dishonesty, corruption, and criminal behaviors as a way to achieve high status quickly. This could explain why some individuals in some governing systems of some nations are often corrupt, and why the governing system of our own nation would sometimes support such individuals for the good and security of our nation, at least in the short run. But, in the long run, are we really doing the right thing?

We humans need to think longer-term and longer-range and thus rethink how we should react to how current life is on Earth. We need to also reduce our ego and rethink what it would be a better indicator of status. For example, a better indicator of status could be how well we are doing in helping mankind fulfill its primary purpose for being here on Earth.

This does not mean we have to go against all the attributes of life on Earth such as those attributes listed above. In my mind, it means we need to be knowledgeable about them and then apply our knowledge in a manner that would make mankind's life on Earth to be as close as possible to be how life is in the spirit world.

A major change in our choice of dominant basis for how we carry out our lives from what is stated above to what is stated below would go a long way in helping us do what we need to do to have mankind fulfill its primary purpose for being here on Earth. Our choice of dominant basis should be:

14.2. Empathy and Compassion for One Another

This choice would by itself make mankind's life on Earth come closer to how life is in the spirit world. More specifically, in the spirit world, every human spirit is a part of one another and where having an ego is only a concept and not a practice. While we are physically separate entities on Earth, this choice would make us spiritually more a part of one another. It would then go on to have us behaving in manners that would continue to make us spiritually increasingly a part of one another.

Chapter Fifteen
Combining Spirituality and Science in Our Pursuit of Advancements of Both Kinds

So far, we have talked about technical advancements as pertaining to how things work in our physical world and spiritual advancements as pertaining to how things work in the spirit world. But, the two worlds are connected together since our physical world is designed and brought into being by the spirit world. Therefore, an interrelationship is bound to exist between our technical advancements and our spiritual advancements.

For example, quantum superposition and entanglement are phenomena that exist in our physical world and that scientists so far have not been able to explain. However, the spiritual model presented in Reference 1 was able to provide a logical and explicit explanation for them by taking into considerations the spiritual part of the phenomena. In other words, the quantum superposition and entanglement phenomena are partly physical and partly spiritual,

Actually, everything that exists or could be expressed in our physical world is partly physical and partly spiritual. It is just that for most things we could ignore either the spiritual part or the physical part and still come up with a reasonably good approximation of reality because either the spiritual part or the physical part is so small that it could be ignored.

The spiritual model in Reference 1 indicates physical and nonphysical universes could exist because it is possible that the spirit world could need one or the other at various times to help restore balance. The spiritual model also indicates that the spirit world is unlikely to design and bring into being a universe that is partly physical and partly nonphysical. That would be more complicated than necessary, and the spirit world being very wise would prefer to keep things as simple as possible.

Besides, a universe partly physical and partly nonphysical is unlikely to serve either its physical purpose or its spiritual purpose very well. It would be like a tool that is designed to perform two very different tasks. It is unlikely to perform either task very well. It is better to have two separate tools, each designed for its own task. Similarly, the spirit world is more likely to design and bring into being two universes, one physical and the other nonphysical if it needs to have both physical and nonphysical experiences in order to restore balance.

We might ask; why did I bring up the notion of a universe being partly physical and partly spiritual as being unlikely? I did it because dark matter and dark energy remains a mystery to scientists as to what they are and what they do. It seems as if the scientists could be considering them to be nonphysical things.

But according to the spiritual model in Reference 1, since our universe is a physical universe, everything that makes up our universe, including the outer space of our universe, must be physical. As discussed earlier in Chapter Eleven, I speculated that the outer space of our physical universe could be made of dark matter and dark energy and that these are physical things. In addition, dark matter and dark energy might also be producing the gravitational force in our universe. It has also been a mystery as to what it is that is producing the gravitational force in our physical universe.

The fact that the outer space could be distorted by gravitation force and that gravitation waves have been observed in outer space indicate outer space must be made of physical matter. While the spiritual model in Reference 1 has not provided a clear explain regarding the role of dark matter and dark energy in making up outer space and gravity, it does suggest such possibilities would be worth exploring.

Chapter 15

These are a couple of examples of how interactions between our pursuit of technical advancements and our pursuit of spiritual advancements could enable us to make discoveries that we might not be able to make if interactions were not considered in our pursuit of both technical advancements and spiritual advancements.

The very first such interaction led to the formulation of the spiritual model presented in Reference 1, and it began with the indisputable notion that:

- **Something, somewhere, somehow knows how to enable our universe and everything in it to exist.**

The "something, somewhere, somehow" is spiritual, and the "our universe and everything in it" is physical.

Additional interactions that played a role in the formulation of the spiritual model include the many analogies that are speculated to exist between our physical world and the spirit world. Such analogies are included in the formulation process because I speculated that the origins of how things work in our physical world must be how things work in the spirit world.

For example, every living thing in our physical world starts out as infants and then gradually grows into adulthood and becomes able to do the things adults do. This suggests that the spirit world itself must have begun as a very small spiritual entity and gradually grows to become something substantial and then be able to do all the amazing things that it does. The spirit world must have learned about this concept of growth from somewhere, somehow, and it is likely to have learned from its own growth experiences.

Speculations such as this were made during the early periods of the formulation process for the spiritual model in Reference 1. Such interactions between the spiritual nature of things and the physical nature of things could also be examples of the interactions between technical advancements and spiritual advancements. More specifically, our technical advancements have progressed to a level that enabled such interactions to go from our technical achievements to our pursuit of spiritual advancements to, in a sense, restarted our pursuit of spiritual advancements. Reference 1 presents a more detailed description of how this could take place.

Chapter Sixteen

More Discoveries and Developments the Spiritual Model Could Explain

16.1. Several Discoveries and Developments Were Presented in Earlier Chapters

Several discoveries and developments that were made after Reference 1 was published are presented in earlier chapters in relation with other topics. Additional ones are presented in this chapter.

16.2. The Reincarnation of Spiritual Entities that Serve as Spirits for Living Things

This topic was discussed in Reference 1 only in a cursory manner without going into much detail as to how such an arrangement naturally came to exist in the spirit world and how it works. To help answer any questions that might have come up this topic is explained in more detail in this chapter.

As indicated in Reference 1, the spirit for a living thing is formed by a spiritual entity that is capable of serving as the spirit of the living thing. Such a spiritual entity would naturally be larger than the spirit for which it is serving as, and in most cases, it is likely to be much larger.

We might ask; why does a spirit need to be formed by a spiritual entity that is larger than the spirit? Why can't a spirit simply be a standalone spiritual entity? Looking back at Reference 1, I could have explained this more thoroughly. I pretty much simply said a spiritual entity needs to be capable of serving as a spirit for a living thing in order for the spirit to be formed. And, it was simply implied that the spiritual entity would be larger than the spirit for which it is serving as.

I also said in Reference 1 that it is the spiritual entity serving as a spirit that reincarnates and that it is not the spirit that reincarnates. This is the key for why a spirit needs a spiritual entity to form it and why the spiritual entity needs to be larger than the spirit. Also, according to the spiritual model in Reference 1, reincarnation is a natural part of how things work in the spirit world. Again, this was simply stated to be the case without fully explaining why.

More specifically, because the spirit world needs to periodically design and bring into being new universes, it will periodically call upon many of the same spiritual entities that are capable of serving as spirits for living things to serve again, this time as spirits for the new and different living things that would reside in a new and different universe.

Each time this happens, the spirit that any one of these spiritual entities would serve as would be different from any spirit it has served as before, because the new universes would be different from any that existed before. For this reason, past spirits would naturally not be called to reincarnate and only the spiritual entities that served as past spirits would be called to reincarnate.

Another possible situation is that such a spiritual entity could be called upon to serve as the spirit for a different living thing that would reside in the same universe in which a living thing that the spiritual entity had served as the spirit previously had resided. The different living thing could be the same species as the previous living thing or it could be a different species. In any case, the living thing would be different from the previous living thing such that its spirit would have to be different from the previous spirit.

Because the spiritual entities that are capable of serving as spirits for living things would be called upon to do so for essentially an unlimited variety of living things, they would have to be quite a bit larger than would be any spirit for which they are serving as. This is because each time they are serving as a spirit they would be employing a different portion of themselves to do so. To be able to cover essentially an unlimited number of variations of spirit configurations they would naturally have to be quite a bit larger than any spirit they would serve as.

16.3. Possible Coincidence that Could Be Perceived as a Sort of Reincarnation of a Spirit

We might then ask: how is it that sometimes an individual seems to be very much like some previous individual and sometimes seems to even know some things the previous individual knew? The answer is that it is possible that the same spiritual entity that served as the spirit for a previous individual happened to be called upon to serve as the spirit for a new individual, and the new individual happens to be of the same race and is somewhat related to the previous individual. In that case the new individual is likely to have instincts that are strongly associated with the previous individual.

These two perhaps unusual coincidences could result in the new individual as being perceived as being sort of a reincarnation of the previous individual in terms of appearance and in terms of the instincts of the new individual. According to the spiritual model in Reference 1, instincts come from the spiritual entity that is serving as the spirit, and in this case the spiritual entity embodies the entire life history of the previous individual.

16.4. The Coincidence in the Preceding Section Could Be Explored with Clones

The coincidence discussed in the preceding section of this chapter could be man-made by producing a clone. A clone is likely to be automatically physically like the original individual. It would be interesting to see if the clone would have instincts that reflect what the original individual knows.

However, if this turns out to be the case, then the discussion presented in the preceding section would be true and would thus lend further confirmation that the spiritual model presented in Reference 1 is correct at least to a large degree.

This is only hypothetical. We are not likely to make human clones.

16.5. Spiritual Commonalities among "Sixth Sense" Phenomena the Spiritual Model Could Explain

The spiritual model was able to provide logical and explicit explanations for quite a few unusual things that are presented in Reference 1. Some additional things the spiritual model could explain are presented in this section of this chapter.

Some individuals and some other living things could "sense" certain things typical humans could not sense, or they could sense certain things much better than typical humans could. We often say such individuals or living things have a "sixth sense". The things they could sense or could sense much better would include: telepathic signals, intuition signals, instinct signals, messages received through cross-learning, messages received in dreams, etc.

Some individuals and some other living things could "see" certain things typical humans could not. We often also say such individuals or living things have a "sixth sense". The term "sixth sense" generally suggests an individual or a living thing has one more major sense than typical humans have and that is why they could "sense" or "see" things typical humans could not.

It turns out that such phenomena are not as mysterious or as complicated as the term "sixth sense" might imply. The spiritual model presented in Reference 1 indicates such phenomena are more natural than mysterious and that every human has the potential for such abilities.

Numerous living things such as dogs, cats, horses, cattle, birds, etc. currently have such abilities as documented in Reference 9 by Rupert Sheldrake. They are likely to have had such abilities all along and had never lost them. It is perhaps because they have not pursued technical advancements as we humans have and therefore have not been distracted to the extent that they would ignore their sixth sense abilities to thus allow them to fade as happened with us humans.

Such evidence suggests we human at one time had such abilities as well and that they are still with us in some dormant state. This indicates it is possible for us to revive such abilities as the spiritual model in Reference 1 suggests. This could be why certain such abilities could be learned when taught by experts. For example, the ability to see and read auras and the ability to do remote viewing are two abilities that various individuals have learned how to do when trained by experts.

The spiritual model in Reference 1 indicates various spiritual commonalities exist among such abilities, and this is why I say such abilities are not as mysterious or as complicated as we might imagine. This also means if we could learn how to acquire or revive one such ability, chances are we could also acquire or revive various other such abilities. This is because they have common ways of functioning. Some examples of such commonalities are as follows:

1. **Telepathy, intuition, instincts, cross-learning, dreams, channeling, and fortune telling:**

    The spiritual commonality that exists among all these abilities is they all function by spiritual signals being issued by one or more spiritual entities and are then received by one or more other spiritual entities.

    In the case of these abilities, the signals are sensed by certain spiritual senses that a spirit of a living things would have. These certain spiritual senses would give a living thing its "sixth sense".

We humans are likely to also have such certain senses, but they are currently dormant. We should be able to revive them as some individuals have done regarding their ability to see and read auras and their ability to do remote viewing.

a. Telepathy: Spiritual signals are issued from the spirit of a living thing and are received and sensed by certain spiritual senses of the spirit of another living thing.

b. Intuition: Spiritual signals are issued from one or more spiritual entities that had served, or are serving, as spirits for living things and are received and sensed by certain spiritual senses of the spirit of a living thing.

c. Instincts: Spiritual signals are issued from the spiritual entity that is serving as the spirit of a living thing and are received and sensed by certain spiritual senses of that very same spirit of a living thing.

d. Cross-learning: This is when members of a species of living things have gone through an experience and thus learned something. Then other members of the same species of living things have also subsequently learned the same thing even though they have not gone through the experience that the first members have gone through. What was learned by the first members got transmitted to the second members. The mechanism by which this works would be the same as how telepathic communication works.

e. Dreams: An individual could get messages in dreams. The sources and mechanisms by how this works could be similar to, or the same as any one or all of the mechanisms described in

Items a, b, c, and d.

I received messages in dreams three times in recent years. In the first time, my oldest sister appeared in my dream quite distraught. She is so level headed she hardly ever gets terribly upset. The next morning, she called on the telephone and was as distraught as she was in my dream. She often checks up on our closest uncle who lives alone. This morning she found him passed away in his apartment.

The second time was four years after I got a new roof. In my dream a roof leak occurred and water was coming down from a doorway. Now, what are the chances a roof leak would occur so soon, and what are the chances water would come down a doorway? Four days later it rained and water stated coming down one of the doorways.

The third time, in my dream an empty wallet was found on the front seat of my car as I was going shopping. Then in the same dream an empty wallet was found on the counter of the store. The next day I went shopping and during check out I discovered I forgot to put my credit card back into my wallet after using it to pay for some repair work done in the house a few days earlier. I have essentially an empty wallet.

f. Channeling: This works the same way telepathic communication works except the spiritual signals are issued from the spirit of a formerly living thing instead of from the spirit of a living thing. The signals are then received and sensed by certain spiritual senses of the spirit of a living thing.

g. Fortune telling: As explained in Reference 1, the spirit world is constantly projecting into possible futures. At any point in time the spirit world would form numerous possible futures based on the present and recent past. This is part of how the spirit world prepares itself for any possible future, just as how any very wise individual would prepare for any possible future.

The one projected future that would come true would depend on what all the living things that exist will do, and this includes what us humans will do.

All the possible futures would exist in the spirit world as spiritual entities, just as anything else in the spirit world would be. These spiritual entities would issue signals. The spirit of a fortune teller could sense these signals with its certain spiritual senses.

Then based on what the individual who wants to know about his or her future would say to the fortune teller, the fortune teller could make a fairly reasonable guess as to what the future would be like for that particular individual. This is because there are more likely than not to be a lot of similarities among all the possible futures when it comes to what could happen next for that particular individual.

However, a rare case could come up in which not a whole lot of similarities would exist among all the possible futures for an individual. In this case the fortune teller might not be able to make an accurate guess as to what would happen next for this individual. This could explain why it is a hit and miss thing for fortune tellers and why there are likely to be more hits than misses.

h. Which of all the possible futures would be the one to come true: Every living thing that exists would have a collection of possible futures that could come true for it. Each living thing would have a different such collection. Then when all such collections are superimposed on top of each other, there will be just one possible future that would apply to all living things. This would be the possible future that will be the one to come true.

We might then ask: does this mean an essentially infinite number of such groups of possible futures would need to be superimposed since there are essentially an infinite number of livings that exist? And, in addition, such a process would need to be repeated at every instant as time passes?

The answer is yes for both questions. But dealing with essentially an infinite number of anything repeatedly at every instant is what the spirit world is always doing. Therefore, it is not a big deal for the spirit world even though it would be an impossible task for us humans.

2. **Seeing and reading auras, remote viewing, what certain animals could see, and seeing or sensing ghosts:**

The spiritual commonality that exists among these abilities is the ability to see certain things in the spirit world much like how we could see things in our physical world. Some individuals and apparently some creatures are in touch with certain ones of their spiritual senses enough to be able to see certain things in the spirit world.

a. Seeing and reading auras: As explained in Reference 1, the spirit and the soul of a living thing contain the entire history of a living things life from its beginning to the present. The living thing exists in our physical world because its spirit enables it to exist. It is the spiritual signals that are issued from all the pieces of knowledge making up the spirit and that are transmitted through the soul that goes from the spirit to the physical body that enables the body to exist in our physical world. These spiritual signals flowing through the soul from the spirit in the spirit world to the physical body in our physical world would produce the aura.

An individual who could see and read aura would be able to see the aura in the spirit world with certain ones of his or her spiritual senses. The image of the aura would appear to that individual as a multicolored glow surrounding the body of the living thing. While the image is sensed in the spirit world, the viewer's mind would superimpose it upon the body in our physical world such that it would appear as surrounding the body in our physical world. The entire history of the living thing's life from its beginning up to the present would be recorded within the pattern of the multicolored aura.

We might ask; how does the entire history get recorded in the spirit and soul? The answer is every experience a living thing goes through generates new pieces of knowledge or gains certain already existing pieces of knowledge. These get added to the spirit of the living thing. Each piece of knowledge issues its own unique spiritual signal. Therefore, the combination of all such signals being issued by the spirit and is flowing through the soul would contain the entire history of all the experiences the living thing has gone through from its beginning up to the present.

Three actual examples of the ability to see and read auras are presented in Chapter Two of Reference 1. I have a coworker friend who could see and read auras. She is able to tell what mood I'm in and if I am tired by looking at my aura, and she would be absolutely correct every time. My wife and I attended a women's health fair. A woman who teaches others how to see and read auras had a booth at the fair, and she offered to give my wife and me a reading. She was 95% correct for both of us as to how we view life. She was even able to identify something about my wife and the cause of it that only my wife and I knew. After these experiences I became convinced auras are real and that they do contain the history of the life of a living thing from its beginning up to the present.

The ability to see and read auras could be learned when trained by experts. The woman at the women's health fair who gave my wife and me a reading is a trainer. She also had with her several of the individuals that she has trained.

b. Remote viewing: This is an ability to see what is going on at a remote location without physically being there. It works regardless of how far that location is or if the remote viewer has ever been there. It is my perception based on the spiritual model presented in Reference 1 that a remote viewer is able to see the spiritual form of that location in the spirit world with certain ones of his or her spiritual senses and is then able to translate what is seen into what it would look like in our physical world.

This ability could be learned when trained by experts. However, based on what I know, this ability does not always work perfectly, especially if the remote viewer is unfamiliar with the place being remotely viewed. Apparently, it helps to at least know what the place looks like, such as having a photo of the place, in order to be sure that the place being tuned into is the right place.

c. What certain animals could see: Cats in particular are known sometimes to behave as if they are able to see things that we humans cannot see. Descriptions of this are given by Rupert Sheldrake in Reference 9 regarding the behavior of various other kinds of animals such as dogs and horses that in my opinion are likely to indicate they too are able to see things we human cannot see. Not every dog or horse would exhibit such behavior.

An example of the kind of behavior we are talking about is that such an animal would know how to go from its home to some distant place that it has been before with the owner, but the animal have gone there in a car in which it stayed on the floor of the car and have not seen all the landmarks of the path from home to that place. Or, the animal might have gotten to that place from some location other than home, and it would still know how to go from home to that place.

Another example given by Rupert Sheldrake is a woman riding

her horse in a forest and got lost. She had no idea how to get back home, so she simply let her horse find the way. The horse behaved as if it knew exactly how to get back home, and it took a path that was not the path on which they took to arrive at their current location. They went through areas where she had to open a gate that they had not gone through before. The horse was able to get them home by taking what appears to be the straightest path to get there.

In my opinion, I think some animals are able to see what is going on in the spirit world as well as what is going on in our physical world. Therefore, even if it is not looking at what is going on in our physical world it will be looking at what is going on in the spirit world.

Therefore, while they might be unfamiliar with how to get from one location to another in our physical world, they would be familiar with how to get there in the spirit world. In other words, some animals are in touch with their certain ones of spiritual senses enough to see things in the spirit world. Rupert Sheldrake, Reference 9, gave several examples of this regarding dogs in addition to the mentioned horse. Some birds are apparently able to do this as well according to Rupert Sheldrake.

d. Seeing or sensing ghosts: A person's soul is partly in the spirit world and partly in our physical world when the person is alive. When a person dies, his or her spirit could choose to let his or her soul remain partly in our physical world and partly in the spirit world, or let his or her soul go completely into the spirit world.

If the soul goes completely into the spirit world, then it is possible for someone who is very in touch with certain ones of his or her spiritual senses to see or sense the soul as a ghost in the spirit world but would interpret the ghost as being seen or sensed in our physical world. In this case only those individuals very in touch with certain ones of their spiritual senses could see or sense the ghost.

If the soul remains partly in our physical world, then it is possible for anyone to see or sense the soul as a ghost in our physical world their major senses.

3. **Out-of-body experience, near death experience, pseudo space travel, pseudo universe travel, and space and universe travel in general:**

The spiritual commonality in these cases is when an individual is asleep or unconscious, the mental ability part of the soul could detach from the physical body while the rest of the soul would remain attached to the physical body. The part that remains attached is the part that enables the physical body to exist and be alive in our physical world. Meanwhile the detached mental ability part could go wandering around either in our physical world or in the spirit world.

a. Out-of-body experience: The detached mental ability part of the soul could go wandering around our physical world or around the spirit world. The effect would be the same since what the detached part sees would be the same whether it is seen in our physical world or in the spirit world as long as the wandering covers only the region that is nearby the physical body.

If what is seen covers regions far away from the body, such as a visit to the moon, then the wandering is done in the spirit world and not in our physical world.

If the detached part goes wandering around in our physical world, then it is more likely for anyone to see it as a ghost in our physical world. If the detached part goes wandering around in the spirit world, then only those individuals who are very in touch with certain ones of their spiritual senses might see it as a ghost in the spirit world but would perceive it as being in our physical world.

b. Near death experience: The soul would completely detach from the physical body for a brief period. This consists of the mental ability portion and portion that enables the basic building materials making up the physical body to form the physical body and to enable the body to be alive.

The basic materials making up the body have their own spirits and souls that remains attached to those materials to enable them to exist in our physical world.

If the spirit decides to have the detached soul of the individual go completely into the spirit world, then the individual would be able to see what it is like to reside in the spirit world in his or her afterlife.

If the spirit decides to let the soul remain partly in our physical world and partly in the spirit world, then the individual could have an out-of-body-like experience, and he or she might be able see the medical team working to revive his or her physical body.

If the body is revived quickly enough then the soul reattaches to the physical body. Any degradation caused by the temporary detachment of the portion of the soul that enables the building materials to form the body and be alive would be negligible.

c. Pseudo space travel: This in one of many possible ways to do space travel by going through the spirit world instead of going through our physical world. Conceivably, we should be able to figure out how to do this after we have figured out well enough how things work in the spirit world through our pursuit of spiritual advancements.

The process of figuring out how to do this is likely to be similar to how we figured out how to do space travel in our physical world once we have figured out well enough how things work in our physical world through our pursuit of technical advancements.

I would expect pseudo space travel to be similar to having an out-of-body experience except several individuals could be involved together, and they would be completely aware of each other's participation. They would have control as to where they are traveling within the spirit world to reach the spiritual forms of the places in our physical world they want to visit.

d. Pseudo universe travel: Pseudo universe travel would be done very much like how pseudo space travel would be done. Universe travel of any kind is possible only if one or more other universes exist along with ours.

On the other hand, if no other universe exists along with ours, then pseudo universe travel could enable us to travel to some form of "nothingness", and this might enable us to gain some knowledge about what nothingness is like before our universe was brought into being.

A major difference between pseudo universe travel and pseudo space travel is that the spiritual senses we human have might not include the kinds of spiritual senses necessary to see or sense completely what are in another universe. Every universe is unique and different, and thus living things residing in them are likely to have unique and different sets of spiritual senses as well as unique and different sets of major senses.

For example, we humans might be able to see another universe in black and white instead of in color or certain things in that universe would be invisible and/or intangible to us.

e. Space and/or universe travel by going through the spirit

world in general: There are likely to be other ways to do space and/or universe travel by going through the spirit world besides doing it in a pseudo manner. For example, all degrees of dimensionality are possible in the spirit world. We are so used to being in a three-dimensional physical world that we would have difficulty envisioning what a universe with more or less than three dimensions would be like.

What if the spirit of our three-dimensional physical world could be temporarily added a fourth dimension such that we could use that fourth dimension to instantaneously travel to anywhere in our universe or to some other universe? In mathematics, transformations could be made to make a complex equation more tractable to be solved. Therefore, adding a fourth dimension temporarily to our three-dimensional world in the spirit world could be similar to making a transformation.

Another possibility is to take advantage of how, for example, every electron in our physical universe exists because a single spiritual entity enables all electrons to exist in our physical universe. The same goes for protons, neutrons, elemental atoms, and every other basic building material making up our physical universe. This means the spirit world is spiritually everywhere in our physical universe. Could we thus enter the spirit world at one location of our physical world and then exit the spirit world at a different location in our physical world and thus do space travel that way?

All such possibilities, and more, might be discovered to be possible if we were to seriously pursue spiritual advancements such that we would become very knowledgeable about the workings of the spirit world. After all, we have discovered amazing things about our physical world that we would have never thought possible until we started seriously pursuing technical advancements and thus become very knowledgeable about the workings of our physical world.

We need to keep in mind that if we are able to discover amazing things about our physical world, we are bound to be able to discover even more amazing things about the spirit world. After all, it is the spirit world that designed and brought into being our physical world in the first place. Therefore, the spirit world is bound to be more amazing than is our physical world.

We also need to keep in mind the spirit world has been designing and bringing into being numerous universes in the past, and that those other universes are bound to be just as amazing as is our particular physical universe. And those other universes are likely to be amazing in ways that are different from how our universe is amazing. Therefore, this indicates that the spirit world is likely to be orders of magnitude more amazing than is our physical world.

Shouldn't we humans be more interested in making discoveries about the spirit world than doing things and behaving in a manner that make a mess on Earth?

4. **Dreams, restoration process, long-term memory, and short-term memory:**

The spiritual commonality here is that the pieces of knowledge that are recently generated or that are recently gained access to are being reviewed or reassessed by the spirit world for a variety of different purposes.

a. Dreams: A common perception is that all the experiences of the day are being organized and put into long-term memory storage. This is then assumed to produce dreams and also to create long-term memory. However, according the spiritual model presented in Reference 1, this perception could be only a secondary purpose of sleep. The primary purpose of sleep is described in the subsection "Restoration process" which immediately follows this subsection.

b. Restoration process: This topic was covered in Chapter Four of this book and in greater detail in Reference 1. It is briefly reviewed here in preparation to address further the topics of long-term memory and short-term memory in the two subsections to follow.

According to the spiritual model presented in Reference 1, we exist in our physical world because signals from our spirit enable us to exist. During the day these signals are drown out by the loader signals created by all the activities and demands of our physical world. With the signals from our spirit being drown out and thus unable to fully maintain our existence in our physical world, our body starts to degrade,

Therefore, we are designed to need sleep in a quiet and darken place such that the signals created by our physical world would not drown out the signals from our spirit. If the signals from our spirit are able to enable us to exist in our physical world, they would be able to restore our body from all the wear and tear it received during the day.

The first step in the restoration process is for the spirit world to review all of our experiences during the day so as to determine the wear and tear our body received during the day. This reviewing process will create dreams and by virtue of being reviewed, the experiences are thus more easily recalled later in life.

Therefore, the primary process during sleep is restoration, and the secondary process could be perceived as being to form our long-term memory.

c. Long-term memory: This topic was touched upon in Reference 1 in a cursory manner. It is discussed in greater detail here to answer any questions that might come up.

The formation of long-term memory is also covered in the

discussion on restoration. What is covered here is why senior individual's long-term memory is better than their short-term and recent memories. As explained in Reference 1, long-term memory is formed when the individual was younger and did not have all the disabilities that come with old age. Thus, their experiences were well formed and are therefore easy to identify and recall. This enables senior individuals to have good long-term memories.

Their recent experiences are generally not well formed because their dominating ongoing experiences dealing with their disabilities are mixed in with their recent experiences. This renders their recent experiences to not be well formed and therefore not easy to identify to be recalled. Thus, their short-term and recent memories tend to be poorer than their long-term memory.

d. Short-term memory and forgotten experiences: Short-term memory was also touched upon in Reference 1 in a cursory manner. It is discussed in greater detail here to answer any questions that might come up. Added to the discussion here is the subject of forgotten experiences, which has not been discussed in Reference 1.

Recent experiences during the day that have not yet been reviewed by spirit world while we sleep would make up a person's short-term memory.

Usually, such recent experiences would be eventually reviewed by the spirit world in preparation for performing its restoration process. The reviewing might be done at some later date perhaps days and weeks later when the reviewing of such experiences would fit better with the kinds of restoration to be performed at that time.

If such recent experiences never get to be reviewed by the spirit world for whatever reason, they will likely be eventually forgotten, and the degradation formed by such experiences might never to be restored. Most likely such degradation would be minor, and that could be why they inadvertently got overlooked by the spirit world.

What this also suggests is that the restoration process is not likely to perfect every time, if ever. Obviously, this is the case. Otherwise, we would never degrade while we age.

What is said here would thus be consistent with what has been stated about the spirit world very early in this book and in Reference 1, and it is that the spirit world could never be perfect because it could never possess every single piece of knowledge that could be generated. Our being able to eventually forget some of our experiences and that fact that we will degrade as we age would be indications of the spirit world's imperfection and thus its inability to do perfect restorations.

e. Loss of memory after a major surgery: Individuals who had a major surgery could have some memory lost. The spiritual model presented in Reference 1 can explain how this could happen. The experience of a major surgery is so intense that the pieces of knowledge it generates would issue signals that are extremely strong. The memory of the surgery would thus last a long time. In addition, the signals would be so strong that they could drown out the signals of pieces of knowledge that were generated by some recent experiences and also by some experiences that soon followed the surgery. Consequently, such experiences could be difficult to recall or could be forgotten.

I personally know two individual who had this happen. One had open-heart surgery and the other had hip replacement surgery. I did not mention their loss of memory to them since they are bound to have enough on their mind about recovery. Unless the lost memory is very important, it is probably best not to mention it, because he or she is bound to be concerned enough about recovery.

16.6. Matter and Anti-Matter

The conventional thinking is that matter and anti-matter ought to exist in equal amounts in our physical universe. This is based on the assumption that if matter is brought into being from nothing, then to balance things out, an equal amount of anti-matter would be brought into being at the same time. This means galaxies, stars, planets, black holes, etc. made of anti-matter ought to exist in our physical world, and their amounts ought to match those made of matter.

However, according to the spiritual model, a state of balance must exist in the spirit world but not necessarily in any universe the spirit world designs and brings into being. This is because when imbalance occurs in the spirit world, the spirit world would design and bring into being a universe that is imbalanced in the opposite way such that the universe could help restore balance in the spirit world.

This could explain why our physical universe so far appears to be made of matter only and that if any anti-matter happens to exist because of various astronomical events or as a result of certain laboratory experiments, it would only be a tiny amount and would exist for only an instant.

The reason anti-matter could even be produced in our physical universe is because the ability to bring anti-matter into being exists in the spirit world as well as the ability to bring matter into being exists in the spirit world. Both abilities exist in the spirit world in order for the state of knowledge to be balanced in the spirit world.

This suggests that some physical universes would be made only of anti-matter. Such universes are likely to be designed and brought into being as often as those made only of matter. This indicates that when we humans figure out how to do universe travel, we better make sure the universe we want to travel to is not one that happens to be made of anti-matter.

16.7. Artificial Intelligence

Artificial intelligence (AI) is a big thing at the present time and is likely to continue getting bigger. AI enables mankind to do some amazing things and to do some things better than us humans can do. However, it is also currently a major reason jobs are being lost because jobs that could be taken over by automation are increasingly being taken over by automation, and AI is the major reason automation is possible in many cases.

Therefore, in my opinion, it remains to be seen how far automation could go before it becomes a negative factor in the well being of mankind. After all, not everyone could get the education necessary for the more complex jobs that could not be automated, and the number of such jobs might not be enough to keep every able person employed.

AI currently comes in two forms:

1. One form is that we humans do all the thinking for it by programming every situation we can think of into the computer that is behind this form of AI.

2. The other form is that the AI is able to learn on its own on how to handle certain tasks. Therefore, this form of AI is able to do some of the thinking itself.

Some individuals are concerned that AI could someday take over the world and perhaps enslave us humans because AI could become more intelligent than us humans. This cannot happen with the first form of AI. Depending on how an AI is constructed, theoretically the second form of AI could potentially do that if it is also able to develop an ego and a desire to enhance its survivability.

This means as long as we humans make sure AI's could not develop an ego or a desire to enhance its survivability, then, I think AI's would not develop the desire to take over the world and enslave us humans. In other words, don't make AI's to have the same values and attitudes we humans have.

There is a danger that we humans could inadvertently make AI's have our values and attitudes. For example, AI's that are could do facial recognition has proven to be capable of being bias against certain racial groups.

Chapter Seventeen

The Spirits and Minds of Young People Are Our Greatest Treasure for They Are the Future of Mankind

17.1. Helping Young People Prepare for Their Roles in the Future

An important part of assuring a good future is helping our young people prepare for their roles in the future. We are not likely to know what the future will be like. Therefore, helping them prepare for the future does not mean preparing them to handle a specific future situation. What it means is to help them be able to think critically such that they would be able to figure out what they should do and what would be the outcome of what they do no matter the future situation they will face.

Therefore, I think one of the main things we can help our young people to prepare is to help them develop critical thinking skills. Sometimes trying to do this feels like an uphill competition with all the other demands we put on ourselves and on our young people in today's super busy way of life. As soon as our children are old enough to start school we fill their schedule, and ours, with nonstop activities partly because we think it would give them a better chance at being accepted by major universities and partly because we are competing with others who are doing the same with their children.

This means once our children start school, not much time is left for us to help them develop critical thinking skills. The typical school curriculum is filled with so much material demanded to be included that development of critical thinking skill is not necessarily a part of what is taught. This situation typically continues all the way through high school.

So, it seems the only time we have available to focus on developing critical thinking skills is before our children start school. This period would typically be before our children reach age four or five. This then makes me wonder if at ages younger than four might be too young to try helping our children develop critical thinking skills. What we need are some experts on children learning ability at various ages to help us determine what age is the most effective and what age is the least effective in helping children develop critical thinking skills.

17.2. What Are Critical Thinking Skills?

I suspect every individual will have his or her personal view as to what are critical thinking skills. The following is what I think critical thinking skills ought to include:

1. Maintain a positive outlook and think in terms of possibilities.

2. An ability to assess situations, values, and priorities.

3. Understand there are short-term short-range ways of thinking and doing things, and there are long-term long-range ways of thinking and doing things.

    Realize that both could be correct and yet both could be mutually opposing.

    The better one to follow would depend on the situation, values, and priorities.

    But keep in mind we are not likely to have a good future unless what we choose to do would be such that a good future is likely assured.

4. An ability to be imaginative and creative. I think being creative is one of the most important abilities we could have.

5. Realize anything reasonably formulated would be correct but incomplete instead of declaring something as simple wrong.

   Realize there are always two or more sides to any issue.

   Avoid a win-lose outcome. Go for a win-win outcome whenever possible.

   Therefore, keep the interaction and communication open.

6. Have numerous abilities to extend what is known to help understand something that is an unknown. The following are examples:

   a. The ability to use analogies.

   b. The ability to extrapolate forward, backwards, and sideways.

   c. The ability to recognize "missing pieces of puzzles" and to interpolate to find and try out possible "pieces" that could fit and thus "solve the puzzles".

   d. The ability to do unconventional thinking and to explore unconventional concepts

7. Think logically and rationally. Take into account the emotional state of someone, but stay logic and rational.

8. Be aware of the current state of human nature and its effect on the way some people think. Take such effects into account in your own thinking. For example:

a. Some people might argue that two "wrongs" could make a "right" when it just means we have two "wrongs".

b. Some people would not answer a question directly but would go around it or would use the opportunity to promote something else. Just because they say something doesn't mean they answered the question. Politicians tend to do this.

c. Some people would take credit when things go well and blame other when things go badly whether it matches reality for not.

9. Own up to what is yours and not blame others.

10. Be interested in, and be aware of, a wide range of things and topics, especially those having to do with current affairs, how things work, mankind's human nature, and the possibility of politics in various human interactions.

Part Three

Things that Could Become Commonplace in the Future

Chapter Eighteen: Studies and Explorations for Getting More In Touch with Our Spiritual Senses

Chapter Nineteen: Things We Could Sense in the Spirit World

Chapter Twenty: Things We Are Likely Able to do by Going Through the Spirit World

Chapter Twenty One: Will Jesus Come Again?

Chapter Twenty Two: Mankind's Potential Future Way of Life in Partnership with the Spirit World

Chapter Eighteen

Studies and Explorations for Getting More in Touch with Our Spiritual Senses

18.1. Studies and Explorations to Help Us Learn How to Get More in Touch with Our Spiritual Senses

We humans are currently in touch with our spiritual senses enough to find thoughts, intentions, solutions, concepts, designs, and have instincts, intuition, imagination, creativity, etc. This would be enough to enable us to pursue spiritual advancements to a large degree. However, to do it to the degree comparable with the degree to which we are able to pursue technical advancements we need to be more in touch with our spiritual senses.

The notion that living things have spiritual senses that function in the spirit world is first introduced in Reference 1. This makes sense because, according to the spiritual model presented in Reference 1, we are carrying out our lives simultaneously in our physical world and in the spirit world, as also explained earlier in this book. The notion that we have spiritual senses that function in the spirit world could explain the things we could do as described in the preceding paragraph.

The reason we are able to pursue technical advancements to the amazing degree we have done is because we are very in touch with our five major senses and are therefore able to sense very well the workings in our physical world. Ideally, we would like to be as in touch with our spiritual senses as we are with our five major senses. We probably cannot achieve that, but it would be good to at least be more in touch with our spiritual senses than we are now.

A lot of information is presented by Rupert Sheldrake in Reference 9 that, in my mind, could be perceived as evidence that we humans could be more in touch with our spiritual senses. The following are examples:

1. People who have been to Africa often tell stories about the way some Africans can anticipate arrivals in the absence of any known means of communication. For example, Bushmen in the Kalahari Desert of South Africa would know telepathically when members have gone hunting fifty miles away whether they were successful in their hunt before they return home.

   According to the spiritual model in Reference 1, these people are in touch with their spiritual senses enough to be able to communicate telepathically. This means the rest of us humans have the potential of being in touch with our spiritual senses enough to be able to communicate telepathically as well.

2. Some animals and birds are able to find their way back home or to some other place without having traveled to or known the path that they took to reach those places. If we take into consideration that some animals, particularly cats, sometimes behave as if they at looking at something that we cannot see, in my opinion they might be looking at something in the spirit world that they are able to see with their spiritual senses. After all, people who could see and read auras are, in my opinion, able to see the aura in the spirit world with their spiritual senses.

   Since we humans are basically animals, we should be capable of also being in touch enough with our spiritual senses to be able to see things in the spirit world as well.

3. I think some individuals are able to do the following things be cause they are naturally, or are trained to be, in touch with their spiritual senses enough to be able to see the things listed below with their spiritual senses:

a. See and read auras.

b. Do remote viewing.

c. Have out-of-body experiences at will and is thus able to see things without the use of their eyes.

d. See or sense a ghost no one else could see or sense,

e. Sense something is hidden at a specific location without being able to see it there or to know ahead of time that something could be there.

These examples suggest that we humans in general are capable of learning how to be in touch with our spiritual senses enough to be able to do such things and more.

The following are some studies and explorations that could help us learn how to become more in touch with our spiritual senses. This list is only the beginning. As we learn more about what is possible, we are likely to come up with other and more effective ways to become more in touch with our spiritual senses.

1. See if individuals who are already in touch enough with their spiritual senses to see and read auras could easily learn how to apply their spiritual senses to do additional things such as doing remote viewing, having out-of-body experiences at will, seeing ghosts, etc.

2. See if we could confirm that animals such as cats, dogs, horses, cattle, and bird could see things in the spirit world, and whether this plays a role in their ability to find their way to certain places even though they have not seen how to get to those places before.

3. In the case of pets that could sense their owner's thoughts, they

somehow have formed a telepathic communication path with their owners. However, in general the owners have not been able to sense their pet's thoughts. See if such owners could be trained to sense their pet's thoughts.

4. Stories have been written about people being able to sense the thoughts of animals. See if such people actually exist, and if they exist, see if we could figure out how they do it.

5. Some indigenous people in some nations could communicate telepathically. See if we could figure out how they do that. Do they feel it or hear it or sense it in some other way?

6. See if such indigenous people sense the thoughts of people who are not members of their group. Also, see if they could sense the thoughts of people who cannot communicate telepathically.

7. See if such indigenous people could sense the thoughts of animals.

8. For individuals who could see and read auras, see if they could feel any difference between when they are looking at the aura vs. when they are looking at the person. This might indicate if they are seeing the aura in the spirit world instead of in our physical world.

9. For individuals who could see and read auras, can they see auras surrounding animals, birds, plants, nonliving things, etc.

In Reference 1, my friend Rachel saw a white aura surrounding my mom's car and me as I was putting a car cover over the car. This suggests she could see auras surrounding nonliving things under certain situations such as a prized object of a recently deceased person. It would have been interesting to see if she could see an aura surrounding the car without my being next to it.

See if others who could see and read auras are able to see an aura surrounding a nonliving thing such as a car with and without someone next to it, and whether that nonliving thing has to belong to a recently deceased person who really treasured that nonliving thing.

10. See if individuals who could strongly sense intuition and/or instincts are more able to develop abilities to do things such as see and read auras, do remote viewing, have out-of-body experiences, see ghosts, etc.

11. Animals in general seem to have stronger instincts than humans have. Figure out if and how this is associated with how they are able to find their way home or to various other places. Also, see if this is associated with their ability to see things in the spirit world.

12. Some cats and dogs are said to pick their owners. In other words, they seem to know they like a person the minute they see the person. See if they are able to see a person's aura and are in a sense reading the aura. As discussed earlier a person's aura contains the history of the person's life from its beginning to the present.

13. Find out how to read auras and figure out how this relates to spirituality. Do the same for other kinds of abilities that involves being very in touch with spiritual senses.

14. A black aura is said to indicate the person will soon become deceased whether by illness, cancer, accident, etc. This means auras in a sense could predict the future in very narrow ways. Find out how this works. Also find out if a black aura contains a history of the person's life just as a regular aura does.

15. Try to find out if living things, including humans, who are very in touch with their spiritual senses could sense things in the

spirit world that only pertain to our physical world or whether their spiritual senses could also sense things in the spirit world that pertain to other universes.

16. Find out what is going on physically, mentally, emotionally, etc. with individuals and other living things, especially various animals, who are very in touch with their spiritual senses when they are doing things that involve being very in touch with their spiritual senses.

17. Figure out how individuals who teach or train others to see and read auras or to do remote viewing go about doing their teachings or trainings.

18. Figure out how do individuals who have out-of-body experiences at will go about doing it.

19. Find out what do individuals such as Rupert Sheldrake, the author of Reference 9, think regarding how we humans could become more in touch with our spiritual senses.

18.2. It Might Not Be Difficult to Get More in Touch with Our Spiritual Senses

Based on the discussion in this chapter so far it might seem getting in touch with our spiritual senses enough to be able to pursue spiritual advancements to the degree we have been able to pursue technical advancements is quite difficult. But, in my mind, I do not see it as being as difficult as some of us might think.

Through thousands of years we have been used to seeing paranormal things as something to be scared about and something that could be evil. This fear and perception are two of the many things holding us back from being of the mind to seriously pursue spiritual advancements. My experiences with paranormal things have been very positive, spiritually enlightening, and nothing to fear.

For example, as described in Chapter Two of Reference 1, my siblings and I experienced paranormal events after our parents passed away. Since such experiences are very personal, I described only mine in Reference 1. Four strange events are described and every one was emotional but joyful, and above all, spiritually enlightening. Numerous similar events also occurred after my wife passed away, and they are still occurring now and then more than ten years later. All such events are filled with expressions of love whether they occur after our parents passed away or after my wife pass away.

When I was caring for my wife as she slowly dies, and my own cancer was returning, I said I wondered who will take care of me. She said in a very determined voice, "I will be there". I know her, when she says something in that manner, she really means it. And, she kept her word.

I say we humans are not so far away from getting in touch with our spiritual senses enough to be able to seriously pursue spiritual advancements. After all, numerous individuals are already able to be in touch with their spiritual senses either naturally or through training to be able to see or sense things with their spiritual senses. Numerous other living things, particular animals, behave in a manner that clearly indicate they are able to see or sense things that are not visible to most of us humans. Such ability could only mean they are seeing or sensing such things with their spiritual senses.

My conclusion is if we humans could shed our fear and our long-held perceptions about paranormal things and instead accept such things as being normal, natural, and positive we would find our getting more in touch with our spiritual senses to be not as difficult as we think.

18.3. Interactions and Interplays Between Advancements and Applications Will Accelerate Advancements

A parallel naturally exists between our pursuit of technical advancements and our pursuit of spiritual advancements, because we carry out our lives simultaneously in our physical world and the spirit world. However, our pursuit of spiritual advancements will be more complex than our pursuit of technical advancements. And, it promises to be more fascinating, broader, and more consciousness enhancing than our technical advancements.

The things that make our pursuit of spiritual advancements more complex than our pursuit of technical advancements are the following:

1. The spirit world is bound to be many orders of magnitude larger than our physical world.

2. Therefore, the spirit world is also bound to be orders of magnitude more complex than our physical world.

3. Everything about our universe and every other universe has its origins in the spirit world. Therefore, by understanding more about the spirit world we are likely to also understand more about our physical world in ways we could not imagine.

4. We would also naturally want to work on becoming more in touch with our spiritual senses as we go about pursuing spiritual advancements. By comparison we do not have to work on being more in touch with our major senses as we go about pursuing technical advancements.

We are able to make amazing progress in our pursuit of technical advancements because there are close interaction and interplay between the technical advancements we achieve and our immediate applications of the technical advancements, whether the application are good ones or bad ones. This interaction and interplay accelerate the progress of both our technical advancements and their applications. This is both a good thing and a bad thing. It is good if the applications are good. It is bad if the application is bad, especially when the bad applications complicate and add to the mess we already made on Earth.

Similar interactions and interplays are bound to take place with our pursuit of spiritual advancements and their applications. There will also be interactions and interplays between our spiritual advancements and our technical advancements. In my mind, because what we will learn about the spirit world would help mankind behave better, these interactions and interplays will likely inspire and motivate mankind to work toward fulfilling its primary purpose for being here on Earth.

This speculation is based on the likelihood that bad applications of our spiritual advancements are not possible in the spirit world. This is because everything in the spirit world is directly or indirectly a part of everything else. Any bad application by any one thing would impact every other thing directly or indirectly as well as that particular thing. Therefore, there is an automatic deterrent for any one thing to do a bad application. Also, the feelings of oneness, love, empathy, and compassion naturally pervade the spirit world such that the will to do bad applications would not exist.

Chapter Nineteen
Things We Could Sense in the Spirit World

19.1. Things that Reside in the Spirit World

The following are speculations on what we are likely to see that are things residing in the spirit world based on the spiritual model presented in Reference 1. While the spiritual model achieved a high level of confirmation based on common everyday experiences and observations in our physical world, the spiritual model is at best correct but incomplete, just as the spiritual model itself says about anything that is reasonably formulated.

Therefore, the following things are not going to be totally complete or absolutely accurate. However, they are likely to be correct to some degree. We are likely to be able to see the following in the spirit world:

1. What a piece of knowledge, a connection of the first kind, a connection of second kind, and a spiritual entity would look like.

2. What the 4Qs look like and what a piece of knowledge it could generate and the connection that piece of knowledge would have with the 4Qs would look like.

3. What it is like to be in a place that is dimensionality different from three-dimensions.

4. What it is like to be in a nonphysical world.

5. How it is that everything is directly or indirectly a part of every thing else as predicted by the spiritual model in Reference 1.

6. Spirits and souls of human, of other living things, and of nonliving things.

7. How it feels to be in a place in which the feeling of oneness, love, empathy, compassion, pervades the entire place. Some individuals who have had a near death experience have already experienced this.

8. How we would be a part of everything the spirit world does after we have passed away on Earth and our afterlife in the spirit world continues.

9. How the spirit world grows as new pieces of knowledge are generated and added to the spirit world.

10. The enormous range of living things possible.

11. Whether or not all creatures regardless of which universe they reside in would be constructed similar to how all creatures that reside on Earth are constructed.

19.2. The Origins of Things that Exist in Our Physical World

The following are things in the spirit world that enable things to exist in our physical world, according to the spiritual model presented in Reference 1. We will be able to see the following:

1. How the spirit of a living thing goes about piloting the physical body of the living thing residing in our physical world.

2. The spirit and soul of a small black hole and of a giant black hole.

3. What it is going on in Earth's core. There is supposed to be a "nuclear furnace" going on in there.

4. What is going on deep inside of the large planets of our solar system such as Jupiter and Saturn?

5. The role of dark matter and dark energy in the makeup of our physical universe.

6. If the outer space of our universe is a physical thing, what is it made of? Is it made of dark matter and dark energy as speculated in Reference 1 and in this book?

7. The edge of our physical universe and what is beyond the edge.

8. How correct is the spiritual model presented in Reference 1? Do the 4Qs actually exist?

9. How our universe looks now. What we are able to see from Earth is how our universe looks like billions of light-years ago.

10. What are the causes that could make the state of knowledge in the spirit world to go out of balance?

Chapter Twenty

Things We Are Likely Able to do by Going Through the Spirit World

The following speculations are based on various observations and some logical assumptions drawn from the observations. Some of such observations are presented in this book and more are presented in Reference 1. An example of such observations is that the spirit and the soul of a living thing contain the entire history of the living thing from its beginning to the present as indicated by what could be seen and read in the aura of the living thing.

1. Learn about our role in the spirit world such as:

   a. Learn how it is each of us is a portion of the spirit world and that what we do with our life on Earth has an effect on the state of being of the spirit world.

   b. Learn how it is that we have a part in determining what the spirit world does and decides to do.

   c. We will find out that there is no supreme being in the spirit world and that each perception of God is only a certain portion of the spirit world just as each of us is only a portion of the spirit world.

2. Learn about the entire history of the spirit world including what initiated it.

3. Learn the reason each universe was designed and brought into

being, its purpose, and how well did each fulfill its purpose.

4. Visit all past and existing universes and learn everything there is to learn about each.

5. Visit other planets in our universe and learn everything there is to learn about them.

6. Find out if evolution actually spans over multiple universes as indicated in Reference 1 and in this book.

7. Learn about all the other highly intelligent living things that exist in the spirit world in addition to us humans. Find out where we humans stand relative to them in terms of level of intelligence and other mental abilities. Be ready to be embarrassed, but on the other hand don't be surprised in view of the mess we humans made on Earth.

8. Find out if the Big Bang actually occurred or if our universe is being continuously brought into being such that the process is still ongoing.

9. Find out why our universe is so huge with essentially an infinite number of astronomical things existing within what seems to be an endless outer space.

10. Find out what UFOs and UAPs are and if they are visitors from other planets in our universe or they are visitors from other universes or they are something else.

11. Find out if the spiritual model presented in Reference 1 is correct or not, and if it is correct, how correct is it, and what is not covered by it.

12. Find out if possible if our universe will someday end, how it would end, and when will it end.

13. Nothingness is assumed to exist before the spirit world came into being. Find out if this is true. Learn what the nothingness could be and what it is like.

14. The spiritual model in Reference 1 indicates that the outer space in our universe is a physical thing since it is a part of our physical universe. Find out if this is true, and find out what it is made of if it is a physical thing.

This list could go on and on, and it only touches the surface as to the enormous number of things we could do by going through the spirit world. If the list of things we could do in our physical world is enormously long, the list would be orders of magnitude longer for the things we could do by going through the spirit world.

Chapter Twenty One
Will Jesus Come Again?

21.1. Mankind Has Not Made a Concerted Effort to Evolve Further

As mankind evolves mentally from its beginning to its present and beyond, mankind goes from depending on the spirit world to do the thinking for it to its doing essentially all of its thinking for itself. At the present I think mankind is somewhere in between these two stages. In other words, I think mankind has not yet evolved to where it is doing completely all of its thinking for itself.

In my opinion, the parallel between maturation and mental evolution applies here. As discussed in Chapter Twelve, as we go from being a baby to becoming an adult, we go from depending on our parents to do everything for us, including thinking for us, to adulthood where we would think almost completely for ourselves. At some point along the way in our maturation process we go from fearing what our parents would do if we misbehave to where we are essentially working in partnership with our parents while thinking for ourselves. We have a natural partnership with our parents, and we do what is necessary to make this partnership a mutually benefiting thing.

Regarding where mankind is in its evolutionary process it seems to be stuck at where it generally think being a "God fearing person" is a good thing. Mankind has thus not yet evolved to where it thinks "working in partnership with God, or with the spirit world, (God is just a portion of the spirit world)" is a better way to go. Mankind has a natural partnership with God, or with the spirit world, and it needs to make this partnership a mutually benefiting thing.

Relying on fear to get ourselves to avoid behaving badly, and to thus behave well, is sort of childish because mankind is in essence still relying on God, or the spirit world, to do some of the thinking for it; e.g., to define for it what is bad and what is good. It is also "worldly" because this is essentially what mankind is doing on Earth; e.g., mankind is depending a lot on laws and rules to define what is bad and what is OK.

If we thing about it, it is kind of childish when a lot of us are spending part of our time looking for ways to break laws and getting away with it or to get around laws and rules just as how a child would look for ways to not follow rules set by parents and get away with it. It is especially bad when white collar crimes are committed by powerful people who have ways of hiding their crimes or who have surrounded themselves with people who would shield them from being indicted if caught.

It is much better to be self-inspired and self-motivated by our desire to think in terms of working in partnership with God, or the spirit world, to make the natural partnership we have with God, or the spirit world, be mutually benefiting. We would then be defining for ourselves what constitutes bad behavior and what constitutes good behavior. We need to evolve to where we are at this stage of our evolutionary process such that we are doing all our thinking ourselves.

We might ask: what is holding mankind up from evolving to the stage where it would be doing all of its thinking for itself? The answer is because mankind seems to prefer to rely on fear to get itself to behave well. It is easier than making a concerted effort to improve its overall behavior such that mankind would evolve further.

In my opinion, mankind has been behaving poorly for such a long time that it forgot it has a primary purpose to fulfill for being here on Earth. Therefore, it is now living life for the sake of its rather brief life on Earth instead of for the sake of fulfilling its primary purpose and for the everlasting afterlife in the spirit world.

21.2. The Problem Is Mankind Got Trapped by the Three Major Obstacles that Are a Natural Part of Our Physical World

As indicated in Reference 1, our particular physical world has three major obstacles we are supposed to learn to work with or work around and not get trapped by them. The obstacles are:

> **Obstacle (1):** Our universe is made of materials and energies that can be possessed, controlled, and/or consumed.
>
> **Obstacle (2):** Highly intelligent living things such as us humans on Earth have egos and a desire to enhance survivability.
>
> **Obstacle (3):** Mobile intelligent living things (those that can walk, swim, and/or fly) need to eat other living things to survive.

These obstacles tend to have us take on win-lose competitions with other living things such that our species could survive the conditions our species faced during the early periods of its existence. But we got caught by the trap that are a part of these obstacles, and thus we took on win-lose competitions with other members of our own species in addition with other living things. We seem to have never made a concerted effort to evolve beyond this mentality.

When we have evolved to where we are able to develop ways to no longer need to continue having win-lose competitions with other living things, we would continue to have win-lose competitions with other members of our own species. It is as if we got addicted to having win-lose competitions such that we would look for ways to continue having win-lose competitions with just about anything and anyone.

We need to evolve beyond this stage. To start, we need to make a concerted effort to treat other members of our own species with empathy, compassion, and respect. While mankind is evolving well in the technical part of its mental evolutionary process, it is not evolving well in the spiritual part of its mental evolutionary process.

21.3. What Does "Being Next to Jesus" Really Means?

The various religions would essentially say that Jesus will "return" when we humans have mentally evolved to where we would do all of our thinking ourselves and to thus learn to behave overall well. However, I don't think it means Jesus would actually return. Instead I think it means we will become spiritually close to Jesus when we learned to think for ourselves and have fulfilled our primary purpose for being here on Earth.

During the early period of our evolution, the spirit world sent Moses to give us some explicit guidelines on how to behave well. The spirit world was clearly doing much of our thinking for us during that period. Later as we have evolved further, the spirit world sent Jesus to give us less restrictive guidelines on how to behave well. By that time we humans had evolved to where we were doing more of our thinking for ourselves, and thus the spirit world loosened the guidelines on how to behave well so as to allow us to think more for ourselves. Unfortunately, since then, we humans have not evolved much further in terms of the spiritual part of our mental evolution.

It is my opinion that the notion that "Jesus will return" does not mean Jesus is coming back to Earth. I think what it really means it is something will happen when we humans have evolved to where we are doing all of our thinking ourselves and also that we have fulfilled our primary purpose for being here on Earth, and it would not be Jesus actually returning to Earth.

One version I heard when I was very young was that Jesus would return and take back with him those who are worthy and they would then sit next to Jesus. I think "sit next to Jesus" is closer to being what will happen rather than Jesus actually returning to Earth.

When we humans have evolved to where we do all of our thinking for ourselves and we have fulfilled our primary purpose for being here on Earth, our spirit would have grown with all the new pieces of knowledge we have generated by all the experiences we had to go through to have done what we accomplished. The nature of our spirit would then become close to being like the nature of the spirit of Jesus. It is in this sense in my opinion that we would be spiritually situated close to Jesus in the spirit world, or in a sense we will be sort of like "sitting next to Jesus".

This makes more logical sense than to say Jesus will return to Earth. In other words, it makes more logical sense that we would be essentially be going to where Jesus is instead of Jesus coming to us. This makes more sense because it is something that we have to do rather than something that Jesus would have to do. We are the ones who have a primary purpose to fulfill, not Jesus.

Chapter Twenty Two

Mankind's Potential Future Way of Life In Partnership with the Spirit World

22.1. What Mankind Is Doing in the Present

Life in our particular physical universe consists of a part that takes place in the spirit world and a part that takes place in the physical world of our universe. Believe it or not, according to the spiritual model presented in Reference 1, our spiritual part of life is more active than our physical part of life even though our consciousness is tuned more to our physical world than to the spirit world such that we are likely to think our physical part of life is more active than our spiritual part of life.

Some individuals might think the spiritual part of their lives consists of doing things such as attending church and praying to God or Jesus. But according to the spiritual model presented in Reference 1, our spiritual part of life consists of much more and it is constantly active.

We are constantly spiritually active whereas we are not always physically active even if our profession is such that we are constantly physically active during the day. According to the spiritual model in Reference 1, spiritual activities include mental activities such as thinking, learning, imagining, creating, caring, etc. and restoration activities such as healing from injuries, recovering from illnesses, nonstop restoring our body especially effectively when we sleep, etc.

Examples of when we are not very physically active or not physically active at all, but we are spiritually active, would include watching television or a movie, reading a book, conversing with someone, attending a meeting, riding in a vehicle or an airplane, etc. In all such situations we are mentally active, therefore spiritually active. Our spirit is constantly restoring our body from the wear and tear it gets during the day. Restoration takes place most effectively when we sleep, and that is why we need sleep. That is why many nonhuman living things also need sleep.

Because our consciousness is tuned to our physical world more than it is tuned to the spirit world, we are usually unaware of the presence of the spirit world. We might believe something like the spirit world exists, but we hardly think much more than that about it, particularly when we have of lot of other things on our mind.

In addition, almost all of our mental activities have something to do with what we are doing in our physical world. Therefore, we tend to behave as if our physical part of life is what our life is about. It thus follows that when we bring up children, we would focus on helping them learn the following:

1. How to physically survive in our physical world.

2. About the need to maintain a reasonable level of survivability; i.e., a reasonable safety margin to assure physical survival if a crisis should occur.

3. How to function effectively with every living and nonliving thing being separate entities.

4. About the need to use time constructively.

5. How to compete, and to realize some degree of competition is a part of a lot of activities in life.

6. How to think critically.

7. Etc.

There is nothing wrong with these activities. But there are good, neutral, and bad applications of them. With mankind's current state of human nature, bad applications will occur. Thus, mankind's overall behavior will be poor, as how it is presently.

All these activities involve doing things in our physical world. Consequently, the children are likely to think their life consists of only their physical part of life. However, if the children attend church, they are more likely to realize life also has a spiritual part. But their attention is likely to still be focused mainly on the physical part of life, because the spiritual part is likely to be too intangible to fully understand or to clearly describe even for adults, let alone children.

Evidence of this is the fact that we have many different religions and many different denominations of some religions. But now the spiritual model in Reference 1 is able to provide a clear and explicit description of the spiritual part of life. However, this spiritual model is not yet well known, and it may or may not become well known.

Therefore, the children when they become adults are likely to bring up their children more or less the same way their parents did. Consequently, generation after generation, mankind is not likely to achieve much spiritual advancement, even as mankind would achieve amazing technical advancements.

The spiritual model presented in Reference 1 could encourage mankind to pursue spiritual advancements as well as technical advancements by providing a logical and explicit description of the spirit world, how it functions, and how each of us has a natural partnership with it. No other existing spiritual model is able to do this in a manner that is as logical and explicit as the spiritual model in Reference 1 could.

The pursuit of spiritual advancements can inspire and motivate mankind to improve its overall behavior. I think the best way to do this is to start by helping children become more aware of the spiritual part of their lives in addition to being aware of the physical part of their lives. They would thus have a more complete knowledge about life and therefore will be more able to carry out their lives in a more meaningful and more constructive manner. Ultimately, they could avoid making a mess on Earth and could also clean up the mess mankind has made on Earth.

22.2. What Mankind Could Gradually Be Doing in the Future

As indicated at the close of the preceding section of this chapter, in order for mankind to improve its overall behavior, it needs to understand the spirit world more and the natural partnership each individual has with it more. Mankind would thus have a more complete knowledge about life. The spiritual model in Reference 1 is capable of helping mankind achieve this.

Up to now mankind knows a lot about the physical part of life but not much about the spiritual part of life. Any time we attempt to do anything by knowing only half of what we need to know about doing it, we are likely to do it poorly. This has been the case regarding how mankind is carrying out life, and this is why mankind has made a mess on Earth.

As mankind learns more about the spirit world and mankind's spiritual part of life with the help of the spiritual model in Reference 1, mankind could convey the knowledge to children. People develop their values during childhood. Therefore, as indicated in the preceding section of this chapter, a good way to improve mankind's overall behavior is to start by helping children develop values that are based on a more complete understanding of life.

Another important factor to consider is that any significant change in values and behavior needs to be done gradually in order for the change to be lasting and be viable by allowing its progressions to gradually migrate and expand into, and modify, its environment such that the change in values and behavior would be compatible with, and be acceptable in, its environment. I.e., its environment needs to change along with the changes in values and behavior, and environments generally takes a long time to change. This means too rapid a change in values and behavior is likely to end in failure. This could be the reason past and current attempts at improving mankind's overall behavior have not been very effective.

This is, in a backwards sense, analogous to how evolution works. In evolution, the environment changes gradually and a living thing's species has to gradually change with it in order to remain compatible with the changing environment and to thus remain viable as a changed species. If the environment changes too quickly, the species might not be able to change rapidly enough to remain compatible with the changing environment and will therefore go extinct. Any time a species is not compatible with its environment, it is not likely to remain viable.

The situation under discussion is in a sense backward from how evolution works in that it is the values and behavior that are changing, and it is the environment that has to change with them in order for the values and behavior to be lasting and succeed. The analogy is that compatibility needs to be maintained between what is changing and what must follow the change in order for the change to be lasting and succeed.

The spiritual model presented in Reference 1 has a feature that could help speed its rate of migration and expansion into the environment and hopefully modify the environment faster. The feature is that because engineering logic was taken into consideration in its formation, it is capable of providing logical and explicit explanations for all common everyday experiences and observations. This enables the model to feel like a natural part of our lives, and thus its migration and expansion into the environment could take place at a faster rate and thus change the environment at a faster rate.

Humans are creatures of habit. Any sudden change that require a sudden change in habit is likely to be rejected, even if the change could lead to a better future in the long run. This tends to be the case for any change in any area of life. This could also explain why our current fast pace of life is so stressful. We face fast and significant changes constantly in today's high-tech world. A person starting his or her career is likely to have to change jobs and get retrained for a new field of work three times or more over his or her lifetime. Staying in one field of work over a lifetime is now likely to be rare.

How this applies to an effort to improve mankind's behavior is that the effort needs to be gradual. It will take many generations and possibly hundreds of years. Nevertheless, we need to start. Therefore, in addition to the activities listed in the preceding section of this chapter we need to help children understand the following as well, and we need to do it in a manner that children could understand:

1. The spirit world exists. Our physical world and everything in it exist because the spirit world enables them to exist.

2. A description of the spirit world, what it is made of, how it is constructed, and how things work in it.

3. Lives in our physical world consists of a spiritual part and a physical part. The spiritual part lasts forever in the spirit world and the physical part lasts only for a while in our physical world.

4. How each part of life works and how the one part is connected with the other.

5. There is a reason why the spirit world enabled mankind to exist here on Earth. In other words, mankind has a primary purpose for being here on Earth.

6. Mankind is to make its life on Earth to be as close as possible to being how life is in the spirit world. The better mankind is able to do this, the more it would be fulfilling its primary purpose

for being here on Earth.

This means while we are to carrying out our life physically as separate entities, we need to be carrying out our life spiritually as if everyone is a part of one another. This is because while we are separate entities in our physical world, we are all a part of one another in the spirit world.

7. This means, diversity is a natural and valuable feature of mankind. Valuing diversity and applying it in positive and constructive manners would enhance every ability mankind has and would also enhance the quality of the results from mankind's activities. This is because valuing diversity and applying it positively and constructively would greatly broaden the knowledge base mankind could tap into to accomplish just about anything.

8. Etc.

How the listed items work is explained by the spiritual model presented in Reference 1. However, it is only one possible spiritual model capturing only one part of the spirit world. As with all models reasonable formulated by humans, it is correct but incomplete. For this reason, it is my hope that others will be able to formulate additional spiritual models that also take into consideration concepts associated with science and engineering.

The greater the number of such spiritual models, the more complete would be our knowledge of the spirit world. Thus, the more complete would be our knowledge about life. Again, life consists of a spiritual part and a physical part such that both spirituality and science are a part of life. Therefore, ideally, every spiritual model should take science into consideration, and every science model should take spirituality into consideration in order for each to be more complete.

22.3. Mankind Was Likely Quite In-Touch with the Spirit World but in a Limited Way, During Mankind's Early Existence on Earth

In many cases, indigenous people seem to have a greater understanding of their spiritual part of life than most present-day people do. I wonder if they have been visited in their distant past by individuals who are like Moses and Jesus who could provide guidance on good behavior. I also wonder if as mankind begins to emerge from living in caves that the spirit world provided them with guidance on good behavior and knowledge on how their physical part of life relates to their spiritual part of life, and that this was likely done by visitors from outer space and/or other universes.

This could have happened before there were ways to clearly document events, such as how the Dead Sea Scrolls documented some later events. But there does appear to be some artwork on cave walls that could be interpreted as documentations of such visits.

It is thus possible that UFOs and UAP's have visited Earth back then. And, humans at that time might not have the urge to fight or flight when it comes to meeting other humans and other highly intelligent beings. Human over-population was not likely to exist back then such that serious completions would exist among them. Also, it might be that they are quite aware that the spirit world exists and therefore were not surprised by visitors from outer space and/or other universes. Therefore, such visitors might not have had any problems interacting with the humans of those days.

The reason I think humans back then might be aware that the spirit world exists is because some indigenous people and many animals and other creatures of today seem to have extrasensory abilities, as described by Sheldrake in Reference 9. Such people, animals, and creatures have not lost their awareness of the spirit world because they have not been so mentally preoccupied pursuing technical advancements that their awareness of the spirit world has faded to nearly zero due to neglect.

As discussed in Reference 1 and also in this book that highly intelligent living things that are able to learn how to do practical space travel and/or universe travel are likely to be friendly instead of hostile. And, they are more likely to have succeeded in fulfilling their purpose for being where they are. This means they must have a good understanding of the spirit world. If this is true, they are very likely to have visited Earth in our distant past for the purpose of helping the humans of those days to understand the spirit world and the spiritual part of life.

But without a good means to document what they learned, the humans of those days would, generation by generation, gradually forget much, but not all, of their knowledge of the spirit world and of the spiritual part of life. However, they are likely to maintain some of the customs, traditions, and rituals that were developed in an attempt to help them remember what they learned. The customs, traditions, and rituals as well as some artwork on cave walls are what we are able to see today. These might very well be evidence that visitors from outer space and/or other universes have visited Earth back in those days.

References

1. *"Connections with the Spirit World, Second Edition"*, book by Richard Gene.

2. Dr. Marcel Just of the Department of Psychology at the Carnegie Mellon University. Dr. Just was the guest of Lesley Stahl during radio station KCBS's Sunday November 24, 2019, 60 Minutes Radio Program.

3. Luke Kemp, article "The Lifespans of Ancient Civilisations."

4. Bernie Sanders, candidate for president in 2019, made a statement regarding this subject as it pertains to the United States of America in June 2019 and it is paraphrased in the text of this book.

5. Lauren Greenfield, documentary photographer and filmmaker, director of the movie "The Kingmaker" made in August 2019 and numerous other documentaries. Lauren was a guest on the Pat Thurston KGO radio program on December 11, 2019, during which this movie was discussed.

6. Terry Lewis, talks about young people gangs on June 7, 2019 as the guest on the John Rothmann KGO Radio Program.

7. Esther Wojcicki, author of the book *"How to Raise Successful People."*

8. Jamil Zaki, Stanford University professor of psychology, author of the book *"The War for Kindness: Building Empathy in a Fractured World."*

9. Rupert Sheldrake, author of the book *"Dogs That Know When Their Owners Are Coming Home."*

10. Tommy E. Smith Jr., author of the book *"In Spirit and in Truth"*.

About the Author

Looking back, Richard Gene must have wanted to find such a spiritual model all his life. At age nine he pestered the pastor of his church with mischievous questions like, "What if a person sins all his life but just before dying asks the Lord to forgive all his sins; would he go the heaven instead of hell?" The pastor answered "Yes" to this question, and Richard then wondered if he felt uncomfortable when asked such questions. At age 30, Richard's first oil painting was that of a tree with roots showing through the soil, called "The Other Part of Life." The roots signify the part of life not sensed by our five senses; i.e., the spiritual part of life. Richard likes figuring out how things work, and majored in structural mechanics. He earned a Ph.D. at the University of California, Berkeley, California, 1964, under his given name. (Richard Gene is his author name.)

ABOOKS

ALIVE Book Publishing and ALIVE Publishing Group
are imprints of Advanced Publishing LLC,
3200 A Danville Blvd., Suite 204, Alamo, California 94507

Telephone: 925.837.7303
alivebookpublishing.com

www.ingramcontent.com/pod-product-compliance
Lightning Source LLC
Chambersburg PA
CBHW022102150426
43195CB00008B/227